Structural Repair of Traditional Buildings

PATRICK ROBSON
Consultant in Structural Engineering and Dispute Resolution
MSc, CEng, QDR, FConsE, FIStructE, MICE, MaPS

DONHEAD

First published in the United Kingdom in 1999 and reprinted in 2005 by
Donhead Publishing Ltd
Lower Coombe
Donhead St Mary
Shaftesbury
Dorset SP7 9LY
Tel: 01747 828422

ISBN 1 873394 40 3

A CIP catalogue record for this book is available from the British Library

Typeset by Sara Clay, Glastonbury
Printed in Great Britain by TJ International, Padstow

Akaara

£15

Structural Repair of Traditional Buildings

Contents

Preface   vii

Acknowledgements   viii

1.   Introduction   1

PART ONE: BASICS

2.   Structural behaviour   5

3.   Soil behaviour   21

4.   Uncertainty and performance   43

5.   Conservation   53

6.   Health and safety   56

7.   Diagnosis   65

PART TWO: DEFECTS

8.   Causes of structural damage   75

9.   Below-ground defects   80

10.   Ground floor defects   103

11.   Loss of equilibrium   108

12.   Movement   112

13.   Fire   120

14.   Chemical attack on concrete   125

15.   Deterioration   129

PART THREE: OPTIONS

16. Purpose of repair   153

17. Foundations   157

18. Ground floors   174

19. Timber   180

20. Masonry   197

21. Weak materials   221

22. Restraints   227

PART FOUR: MANAGEMENT

23. Preventive maintenance   239

24. Planning and conducting work   245

25. Insured perils   256

26. Law   274

27. Surprises   282

APPENDICES

Appendix A: Risk assessment: health and safety   291

Appendix B: The principles of prevention and protection   293

Glossary   295

Further reading   300

Index   305

Preface

Many different professions are engaged in structural repair, in one role or another – appraising, designing, specifying, adjusting, litigating and, of course, contracting to do the actual work. This book is intended to be of interest to all. It highlights the difference of approach between new design and repair, and it emphasizes the need to diagnose the cause of damage accurately and to consider all the options before specifying a remedy. The problems that have their origins below ground level tend to be the most intractable and to throw up the most extreme options for repair; accordingly more space is given to these than to problems confined to the superstructure.

The purpose of this book is to furnish architects, surveyors, and structural engineers more used to new design, and of course students in these fields, with an appreciation of the technical principles that govern structural repair. When managing serious progressive structural defects, it will often be appropriate to top up appreciation with advice from specialists – structural engineers, surveyors, arboriculturists and contractors who have gained expertise in traditional buildings and their problems.

The book covers traditional materials and methods of construction, principally brick and stone masonry and timber. Concrete is considered only in regard to foundations and ground floors. Repairs to building services and finishes are not discussed; neither are cures for essentially non-structural problems such as damp.

Acknowledgements

The idea and outline for the book was conceived, some time ago, by my colleague Phil Chatfield MCIOB and myself, after we had worked together on many repair contracts. Phil maintained his interest in the book, despite the many other demands on his time, making valuable contributions and checks as the outline was developed. Steve Thomson (Managing Director of SBC (Cambridge) Ltd) applied his practical knowledge to the first draft, and his many suggestions have clarified the text and improved its accuracy.

I am grateful to four people who read parts of the initial draft and offered me their specialist viewpoints: Paul Grundy, of Rogers and Grundy, for advice on surveys; Jeremy Barrell of Barrell Treecare for his arboricultural experience and insight; George Gooden, Senior Partner, Gooden Partnership, for looking at things from the loss adjuster's point of view; and Mike Rowell, Principal of Mike Rowell Associates for once again applying his geotechnical knowledge and experience to my efforts. All their help and encouragement is very much appreciated. None saw the final draft, so I cannot shed any of the blame for anything the reader finds incorrect or displeasing. Finally, my wife managed the book's progress and typed the manuscript and ensured that text and illustrations were assembled on time despite my many stops and starts.

Chapter One

Introduction

There used to be a vogue for obsolescence. Buildings that were not performing well, or not keeping up with our changing needs, were pulled down to make way for something better. This is no longer the case, and most of us hope that the fashion will not return. We now want our buildings to last longer.

When students prepare to qualify in their chosen niche within the industry, their understanding of material behaviour and construction methods is based mainly on new construction. The additional skills needed to maintain, repair and adapt existing buildings are left to the vagaries of experience.

Whereas a new building faces uncertainties that its designer must recognize, an established building has survived them, at least to an observable extent, so that a measure of 'new design' caution can be relaxed. Unfortunately, this is not always understood. In the course of minor repairs or alterations, many old buildings have had their foundations brought up to 'modern standards' at great expense, when the money could have been spent on something more useful. This happens when designers have the confidence to apply only the principles they learned for new design.

On other occasions, repairs have failed because no principles, either new or old, were applied; the symptom was tackled but not the cause.

The perils of over- or under-design can usually be avoided if:

- The cause of damage is correctly diagnosed.
- Its progress (prognosis) is accurately forecast.
- A clear purpose for the repair is established.
- Repair options are analysed.
- Appropriate details are chosen.
- The work is carried out to a satisfactory standard.

The purpose of this book is to provide an appreciation of the problems to be solved on the way to successful structural repair. A basic grasp of building construction is assumed. The gap between basic grasp and appreciation varies from one topic to another and, no doubt, from one reader to another. Some chapters are, therefore, devoted to establishing the principles, structural and non-structural, that shape the technical details. It is hoped that this introductory material will be useful to some readers without being too irritating to others.

Part One discusses basic principles. Chapters 2 and 3 provide, for the non-specialist, a non-mathematical appreciation of the behaviour of structures and soils. Chapter 4 argues that this behaviour is not fully predictable and that there is uncertainty in every design. For new buildings, this uncertainty is codified in British Standards and other orthodox guides, but there is a different frame of reference for existing buildings. Buildings are usually found to perform better than might have been predicted at the drawing board stage. This can be an advantage when it comes to specifying repair. All work has to be reconciled with the needs of conservation, where these apply (Chapter 5) and with health and safety (Chapter 6). Chapter 7 discusses the principles of diagnosis and describes a technique for identifying the true cause or causes of damage by the process of elimination.

Part Two lists the common causes of structural damage to traditional buildings (Chapter 8), and enlarges upon the more important and difficult of these.

Part Three emphasizes the need to be clear about the purpose of repair, and discusses some of the more common options.

Part Four discusses associated topics. Some of the more difficult and costly repairs can be avoided if preventive maintenance is practised (Chapter 23). High levels of intervention, such as refurbishment, impose more formal methods of investigation and design, as discussed briefly in Chapter 24. Managing insurance claims involves additional technical constraints, which are discussed in Chapter 25. Chapter 26 is a very brief reminder of the part the law plays in designing and managing repairs. We conclude (Chapter 27) with a reminder that something unexpected can always be expected.

PART ONE

Basics

Chapter 2

Structural behaviour

An efficient structure transmits the weight of itself and its contents, plus loading imposed by the environment (mainly wind and snow in the UK) safely through itself and into the ground, without undue damage or distortion. Designers use simple models to illustrate the route taken by each load; the route, in turn, defines structural form and determines the internal forces the structure must cope with – or fail.

Figure 2.1 shows the outline of a small building. Loads find their way to supports and from there downwards into the foundations. They do not stop at foundation level, but disperse through the soil, whose behaviour is as important to the health of the building as is the performance of floors, walls and roof.

There is not sufficient detail in Figure 2.1 to make it obvious what happens to the horizontal wind loads. If the building is framed, the connections between horizontal and vertical members (and any sloping members) may be rigid; in this case, the whole frame will respond to loading as a single unit. Modern materials, especially steel and reinforced concrete, have

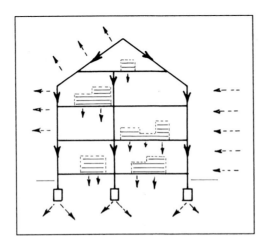

Figure 2.1
Forces on and through a building.

a blend of properties that enables them to enclose large areas and heights within slender frames. Traditional materials are less versatile in this respect, and would not benefit from rigid connections. (The medieval timber frame is a near exception; its performance is enhanced by its connections, but not to the extent of the modern steel or concrete frame, whose connections are crucial to the building's performance.) The external walls of traditional buildings do not gracefully withstand large horizontal loads. Stability must be maintained by spreading such loads through the plane of the floors, and hence into walls that run parallel to their direction (Figure 2.2).

Modelling the structure

The designer needs a reliable means of judging how the proposed or repaired structure will perform when subjected to the various possible loads that will act upon it during its lifetime. It is standard practice to use a stepped method of design, as suggested in Table 2.1.

Table 2.1   Simple structural analysis

Produce a model of the structure	*(Usually a dimensioned sketch, or series of sketches, of the structural skeleton)*
Calculate dead and imposed loads	
Apply the loads to the model	
Calculate forces in structural members	
Calculate stresses in structural members	
Compare stresses with available strength	*(If any stress exceeds material strength, amend member and model)*
Calculate strains (distortions) in structural members	
Compare strains (distortions) with acceptable limits	*(If any strains – distortions – exceed acceptable limits, amend member and model.)*

For simple buildings, outlines such as Figure 2.1, suitably dimensioned, are good enough to serve as structural models.

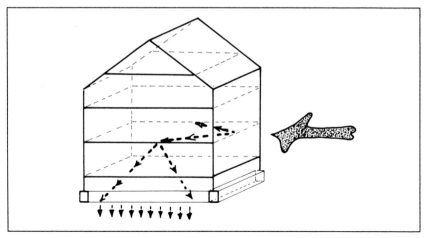

Figure 2.2   Spread of horizontal loading through a building.

Standard loadings

Designers obtain values for the loads shown as unidentified arrows in Figures 2.1 and 2.2 by reference to British Standards. The imposed domestic floor loading, for example, is usually set at 1.5 kN/m^2 . If the floor covering, ceiling boards and joists together weigh 0.5 kN/m^2, the joists must be strong enough to transfer 1.5 (imposed) + 0.5 (dead) = 2 kN/m^2 to the load-bearing walls. Table 2.2 lists the main groups of loading that have to be considered in isolation and in combination, to discover the worst conditions that the building is likely to face.

When commercial buildings are refurbished, owners sometimes require them to be upgraded to support unrealistically high floor loads, so that they can be let for any conceivable use. In practice, even the minimum British Standard loadings are seldom, if ever, exceeded in the lifetime of most buildings (Chapter 4).

Table 2.2   Typical loadings

Gravity: dead	*(self-weight of building)*
Gravity: imposed	*(human occupation, equipment, stored materials)*
Crowd loading	*(where many people can gather, e.g. emergency routes)*
Dynamic loading	*(dancing or jumping; machinery)*
Wind	
Snow	
Volume changes	*(thermal and moisture induced movement)*
Exceptional environmental loads	*(where appropriate : earthquake, mining subsidence, exceptional wind loading, driven snow, dynamic effects)*
Specific hazards	*(associated with building use, e.g. dust explosion)*

Structural form and load paths

Structural form is defined by the way loads pass through the structure.
Figure 2.3 shows some of the forms in common use:

(a) beam and post   (c) truss
      or beam and wall   (d) girder
(b) arch   (e) frame

Truss and girder are, in fact, the same form; the terms are interchangeable, although truss tends to be applied to roofs and girder to other members, such as principal supports spanning openings.

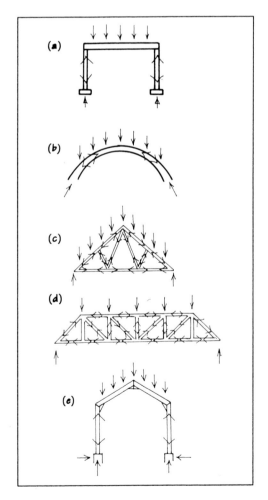

Figure 2.3
Structural form.

Forces in members

Figure 2.4 shows the outline of a simple building consisting of a roof truss and a suspended timber floor supported on brick walls with corbelled footings. As the vertical loads pass downwards, they create forces of one kind or another in every structural member. (For simplicity, only vertical forces are shown here.)

Equilibrium

The first step in structural analysis is to put values to the forces passing through the structure, and to the stresses and strains created by those forces. The starting point is equilibrium.

Loads and forces must always be in equilibrium. This is a natural law that not even the most cunning designer can bend to relieve some practical difficulty. A good designer observes the law and creates an efficient structure that performs well. Poor design commits the building to inefficient means of establishing stability. This can result in unnecessary damage and distortion. If the design is so flawed that equilibrium cannot be achieved at any price, the structure is no longer a structure; it is a mechanism or a disaster, depending on whether your viewpoint is academic or social.

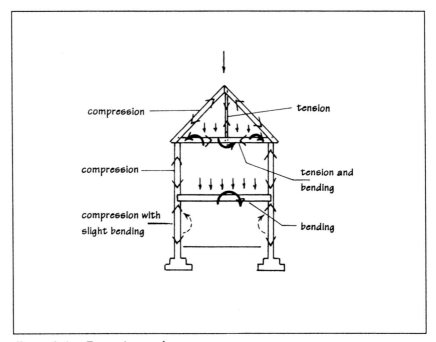

Figure 2.4   Forces in members.

Equilibrium applies both externally and internally. External equilibrium prevents overturning of the structure; internal equilibrium prevents it from breaking apart.

A simple example of external equilibrium is the ladder (Figure 2.5). The single applied load, the person's weight, is equalled by the reaction of the ground (ignoring, for simplicity, the self-weight of the ladder). The propping force, at the top of the ladder, is equalled by the horizontal force at its base. If the ground is too slippery to generate the horizontal force, one of the most common accidents on building sites is initiated. Finally, there has to be rotational equilibrium (better known as moment equilibrium). At either end of the structure (in this case a ladder), the sum of the rotation effect (moments) of external forces must be zero; clockwise and anti-clockwise cancel out.

To calculate equilibrium, the structure (ladder) had to be modelled. Figure 2.5 is the model; and as with all structural models, arbitrary choices were made in forming it. It was assumed that no friction could develop at the top of the ladder. However, if friction were able to develop, it would appear as a vertical force, acting up the face of the wall, and helping to support the ladder (Figure 2.5b). This would change the equations. Alternatively (Figure 2.5c), the ladder might be hooked over the top of

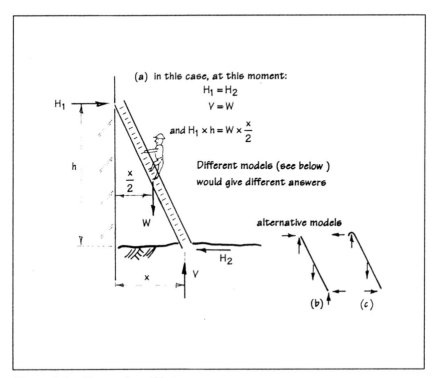

(a) In this case, at this moment:
$$H_1 = H_2$$
$$V = W$$
and $H_1 \times h = W \times \dfrac{x}{2}$

Different models (see below) would give different answers

alternative models

(b)   (c)

Figure 2.5   External equilibrium.

the wall, in which case it would not be dependent for its equilibrium on friction at the base.

In every conceivable case, whatever the detailed model, there will be three conditions of external equilibrium:

- Vertical loads and forces cancel out.
- Horizontal loads and forces cancel out.
- Moments cancel out.

Applying these principles to Figure 2.6, we find that the vertical load at the apex of a simple triangular truss is cancelled by a force of half the value at both supports. The principles of internal equilibrium require the three members of the truss to set up forces within themselves to match (i.e. be in equilibrium with) the load at the apex and the forces at supports. These internal forces can (in this case) be calculated by simple trigonometry.

In Figure 2.6, the internal forces pass in a straight line from end to end of each member. Where this is not possible, the member will experience bending. Bending creates shear forces and bending forces within the bent member, which again are calculated by considering internal equilibrium.

Figure 2.7 (*overleaf*) shows a uniform load on a beam. At every point along its length, the beam has to be strong enough to develop a shear force (a plane force tending to slice vertically through it) high enough to cancel the net vertical loading to one side. It must also be strong enough to develop an internal moment (tending to curl the beam along its length) equal to the net applied moments to one side. Diagrams for these internal forces make characteristic shapes, according to the pattern of loading. The shapes in Figure 2.7 are instantly recognizable as the shear force

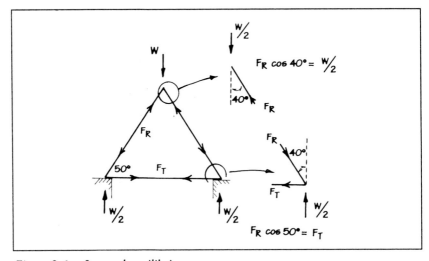

Figure 2.6   Internal equilibrium.

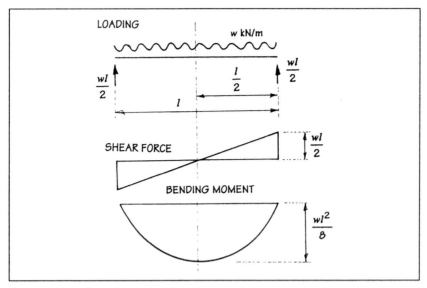

Figure 2.7   Shear forces and bending moment

and bending moment diagrams for a simple beam carrying a uniformly distributed load.

There is one other 'standard shape' worth mentioning. Figure 2.8 shows the shear force and bending moment diagrams for a beam with cantilever ends. Typical features are:

- high shear forces at supports
- a continuous bending moment whose shape depends on the relative size of the load on the cantilever tip.

Figure 2.8 could represent the floor joists in a building whose first floor projects beyond its ground floor, for example, a medieval timber frame.

Stress, strain, strength and stiffness

Having established a working model, with its applied loads and forces in members, the next step in analysing the structure is to calculate stresses. The building modelled in Figure 2.4, illustrating the forces in members, appears again in Figure 2.9a (*page 14*) where it illustrates stresses in members. Methods for calculating these stresses are beyond the scope of this book; suffice it to say that the first test to apply to a structure is that its stresses do not exceed the material strengths. If the maximum tensile stress in the hanger is $1N/mm^2$ and its strength (breaking stress) is $5N/mm^2$, there is a comfortable margin of safety.

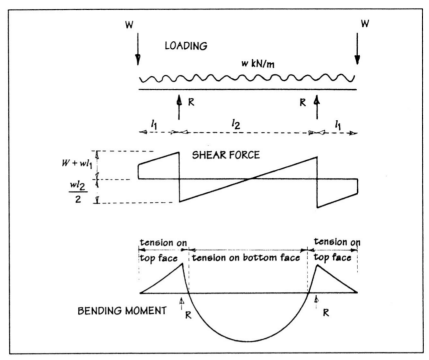

Figure 2.8   Shear force and bending moment: beam with cantilever.

The second test is to check strains (Figure 2.9b). For the purpose of this discussion, strains can be called distortions, consistent with terminology elsewhere in this book. The hanger stretches, the beam bends, and even the walls undergo a certain amount of distortion. These distortions must all stay within acceptable limits. If they become gross, cracks form, openings twist, decorations spall and, although the building remains safe so long as strengths exceed stresses everywhere, it will take on the demeanour of inadequacy. Strains (or distortions) can be derived from the stresses already calculated. The link is the stiffness of the material. Material stiffness is defined by its Young's modulus, also called the modulus of elasticity. (In the case of bending, the geometry of the material – its width and depth – also directly influences stiffness and distortion.) Figure 2.10 (*page 15*) shows typical graphs for Young's modulus. For all practical purposes, there will be a straight-line relationship between stress and strain, within the typical working range. At the top of the line, stress approaches strength, the failing point. The line ends abruptly if the material is brittle; if it is ductile, it flattens out some distance below failure stress.

Although the main purpose of structural design is to ensure that failure is avoided, if failure does occur – because of unforeseen loading, for example – then a ductile failure is preferred to a brittle one, because it gives warning by the large strains (distortions) that occur before the final collapse.

There is one further important facet of strength that ought to be discussed here. If the hanger shown in Figure 2.9 were acting not as a hanger but as a strut (taking the same force numerically, but in compression instead of tension) its stress would be numerically the same, but its strength would *not* necessarily be the same. There is an important difference between tensile and compressive strength. For a given material and grade, the value of tensile strength is more or less the same,

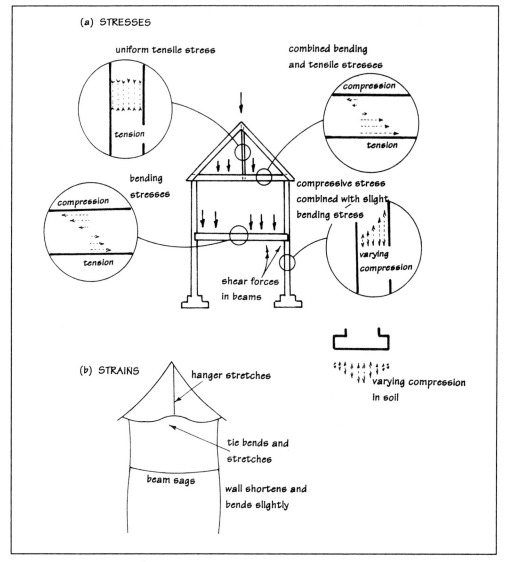

Figure 2.9   Stresses and strains.

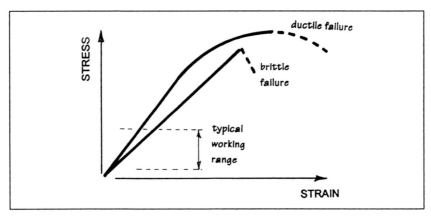

Figure 2.10   Young's modulus.

regardless of the member's length, depth or breadth. However, compressive strength is reduced as it becomes more slender (or long in comparison with section size). At critical combinations of compressive stress and slenderness, the material buckles. For very slender struts, buckling strength can be tiny compared with non-buckling compression strength.

Slenderness dominates the behaviour of many buildings. There is no single definition for the term, because it affects different materials in different ways. For a rectangular joist or beam, the slenderness ratio is length divided by radius of gyration, the latter based purely on the geometry of the element. The larger the beam, the larger the radius of gyration and, therefore, the smaller the slenderness ratio and the greater the compressive strength. Length depends not only on the span between supports, but on how the member rests on or is fixed into each support. Hence, there is an element of judgement in estimating slenderness. A masonry wall's slenderness depends on the distance between vertical supports (cross walls, frames, piers, end walls) and between horizontal supports (floors, edge beams, roof ties). Again, the designer needs to judge the effect of each support. The other factor in determining wall slenderness is its thickness. For a cavity wall, thickness is usually taken as two-thirds of the sum of thicknesses of the two leaves, provided they are well tied. Without ties, they act as independent leaves, each with its own slenderness ratio.

Bending in masonry

It is worth discussing briefly the case of bending in masonry, because it can be an important influence on both appraisal and design. The behaviour of masonry is dominated by the following properties:

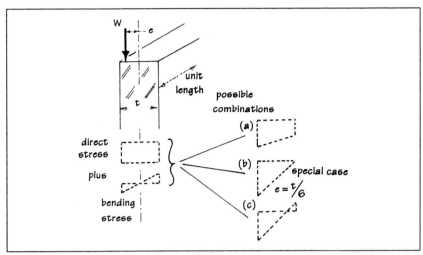

Figure 2.11   Masonry: bending and direct forces.

- It is strong in compression – although this strength is challenged at high slenderness ratios.
- It is weak in tension – most designers make the assumption that plain masonry has no tensile strength; in practice it does have a low but uncertain level of tensile strength.

Figure 2.11 shows how direct compression and bending may combine in the case of a masonry wall. In (a), the uniform direct compression is greater than the maximum bending tension. This is the most common condition. In (b), the two are equal in value, and this creates the unique case of a triangular stress distribution varying from zero to twice the direct compression. Simple arithmetic proves that this happens when the load eccentricity, e, is one-sixth of the wall thickness, t; in other words, the load is at the edge of the 'middle third' of the wall and about to cause tension. In (c), $e > t/6$ and tension forms. If the tensile stress exceeds the critical tensile strength of the masonry, a horizontal crack will form on one face, and some stress redistribution will have to take place to maintain equilibrium. If equilibrium cannot be maintained by this means, however, the wall and anything it supports will topple above the line of the crack.

Figure 2.12 shows a pad foundation with the identical structural problem – identical, that is, apart from one important detail. This time there is no doubt about whether tension (in this case uplift) can develop; it cannot do so with a foundation simply seated on the soil. Once the self-weight of the foundation has been overcome, one edge will simply lift off. To maintain equilibrium, the soil still resisting the loading must form a narrower stress (or pressure) diagram with a larger maximum compressive stress (c), if that can be endured. If it cannot, the foundation will become unstable.

Third and fourth dimensions

For simplicity, structural diagrams are drawn in the plane that best illustrates the principal structural behaviour. It would be dangerous, however, to forget that there are always loads acting perpendicular to that plane, and some form of resistance is needed to maintain the building's equilibrium in that direction. Buildings are said to be robust if the 'perpendicular loads' are resisted comfortably. Figure 2.13 (*overleaf*) shows three outline plans for a masonry building:

- The building in (a) is cellular, with junctions at close centres. It is a strong layout.
- The front and rear walls of (b) consist mainly of openings, so the building has little strength or stiffness in the long direction, unless the openings are framed in steelwork or reinforced concrete.
- The front walls of (c) are more substantial, but they have little restraint except at the ends. Lack of restraint reduces strength, stiffness and robustness.

It is also notable, in the case of (c), that the left-hand end wall can get no restraint from a first floor that is interrupted by a stairwell adjacent to it. Furthermore, the front and rear walls are pierced close to their junctions with it, and are relatively weak at these points. This end wall will, therefore, be more vulnerable to loading in either direction than might be obvious at first sight.

The advantages of regular restraint represented by Figure 2.13a are:

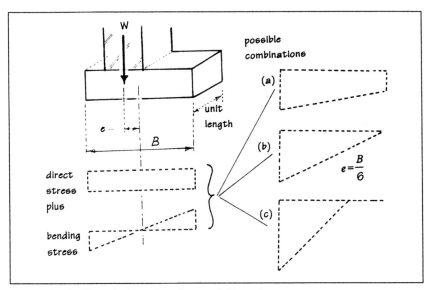

Figure 2.12 Foundations: bending and direct forces.

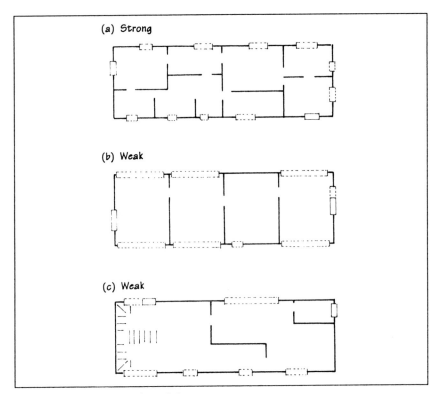

Figure 2.13   Strong and weak layouts.

- It provides ample load paths, especially for horizontal loadings.
- Every wall reduces the slenderness of the walls it joins, so there is a mutual preservation of strength and stiffness.

These advantages can, of course, be diminished by poor bonding (or tying) at junctions, or by lack of integrity within the walls themselves. Examples are: cavity construction with few (or eroded) ties; and walls with rubble interiors that are not well bonded to facings.

If a small building suffers from such flaws, it is readily damaged by the occasional exceptional load and it ages perhaps less gracefully than it might have done. A large building suffers more serious consequences. Temperature changes and other normal loading–unloading cycles cause larger movements in larger buildings, and there can be an insidious accumulation of distortions. Walls have bulged, roofs have buckled and floors have lost support, sometimes suddenly and with serious consequences – all as a result of long-term movement that should have been contained by inherent robustness. Examples are given in Chapters 11 and 12.

Creep

Figure 2.14 shows the typical curve of deflection against time for a timber beam. Initial loading causes an immediate deflection, followed by a delayed action and a long-term creep. If the load is removed, there is an initial recovery followed by a creep recovery, leaving a residual permanent deflection. Even after being dismantled and taken out of the building, timber beams that were previously heavily loaded are left permanently distorted. The effects of creep are seen most often in roof purlins and floor joists that are safe but less than generously proportioned. Purlins sag and give the roof a rippled profile; sagging floor joists may come to bear accidentally, but firmly, on non-load-bearing partitions.

Postscript

One of the repetitious themes of this book is the modelling of structures. A model is needed before structural analysis can begin. It is, in fact, the model that requires engineering judgement; analysis is just arithmetic. The case of joists accidentally bearing on non-load-bearing partitions (mentioned above) is common. It can result in a reduction of the floor's normal deflection, damage to the partitions, or both or neither, depending on the circumstances, which are not entirely foreseeable. That is a trivial example of the uncertainty of structural modelling. Even in simple buildings, the routes taken by principal loading can vary unpredictably.

As noted earlier in this chapter (Figure 2.5, *page 10*), the forces that keep a ladder in equilibrium are alterable by small details that are not always predictable. Hip rafters, spanning from eaves to ridge, are structurally similar to ladders, notwithstanding their shallower pitch. They,

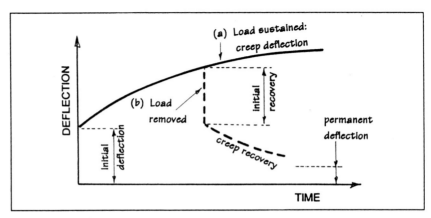

Figure 2.14   Creep.

too, are affected by support details. Thus the performance of even the most simple building elements can sometimes be a matter for debate.

If precision in modelling were vital, this errant behaviour would make structural design a nightmare. Fortunately it is not. The designer need only produce a model that is logical (obeying the principles of equilibrium), safe (ensuring all members and connections have acceptable stresses), and competent (holding distortions within acceptable limits). The building will nearly always perform better than its model, because of accidental enhancement. If the latter fails at any time, for any reason, the structure will at worst default to the more pessimistic design model. (There are occasions – extremely rare in the case of traditional buildings – when this benign axiom does not hold true; there is no escaping the need for engineering judgement.)

The repairer has a different viewpoint from the original designer, because the building already exists. Chapter 4 considers the repairer's viewpoint.

Chapter Three

Soil behaviour

Soil behaviour is influenced by three factors: soil particle size, water content and plasticity. These all interact, and it is their interaction that accounts for most problems with foundations. The three factors are defined below, their interaction is described, and a few of the more important aspects of soil behaviour are introduced. The chapter ends with a brief review of foundation types. (The glossary contains a few explanations of terms used.)

Particle size

For the convenience of verbal description, particle sizes are given familiar terms such as gravel, sand, silt and clay. Table 3.1 (*overleaf*) illustrates their size range, and contains some hints on recognition. A less obvious, but useful, broad division between coarse and fine soils is indicated on Table 3.1. Roughly speaking, it coincides with the boundary between visible and invisible particles. In practice, soils are rarely 'pure' silt or 'pure' clay or whatever, but contain a range of sizes (a particle size distribution). The division between coarse and fine soils is better defined as follows:

- Soils with more than 65% of their particles larger than 0.06mm are coarse.
- Soils with more than 35% of their particles smaller than 0.06mm are fine.

The 65–35 division should only be used as a general guide. Coarse silt and fine sand are on opposite sides of the division, but are difficult to distinguish from each other in appearance or behaviour.

Particle size distribution is often illustrated graphically. Figure 3.1 (*page 23*) shows some typical examples. Fine soil is defined by the shaded area.

Table 3.1   Particle size distribution

Name	Particle size range (mm)	Sub-divisions	(mm)	Field recognition
Boulders	200+	–		} Visually unmistakable
Cobbles	60 – 200	–		
Gravels	2 – 60	Coarse Medium Fine	20 – 60 6 – 20 2 – 6	Gravel is easily recognized and can be classified by tape measure.
Sands	0.06 – 2	Coarse Medium Fine	0.6 – 2 0.2 – 0.6 0.06 – 0.2	Sand always feels gritty. Particles are visible. Unless it is mixed with clay, sand cannot be moulded; unless it is wet, it dusts off easily after handling.
Division between coarse and fine soil. **Rough boundary between visible and invisible particles.**				
Silts	0.002 – 0.06	Coarse Medium Fine	0.02 – 0.06 0.006 – 0.02 0.002 – 0.006	Silt feels silky. Particles are invisible or barely visible. Unless it is mixed with clay, it powders when repeatedly rolled between the palms, but it does not dust off easily. It exhibits dilatency. In water it disintegrates.
Clays	less than 0.002		–	If wet, clay feels sticky and can be moulded. Repeated rolling between the palms produces threads which may be brittle (if non-plastic) or malleable (if plastic). A ball of clay will eventually take a polish. Desiccated clay can be broken into lumps with difficulty, but it does not powder and it is too hard to roll.

Soils with a wide particle size distribution (especially mixtures of fine and coarse)
are difficult to proportion accurately in the field.

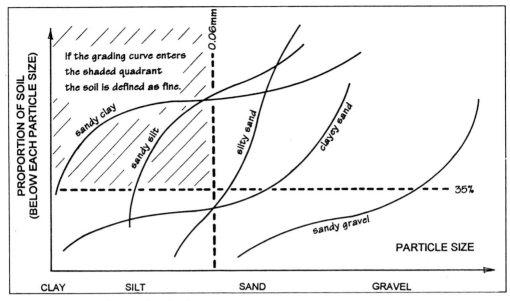

Figure 3.1   Particle size distribution.

Water

Soil is a mixture of solid particles, water and air. Water content (also called moisture content) is defined as the weight of water divided by the dry weight of solid particles. It is usually expressed in test reports as a percentage, implying a proportion of the whole, although it is really a ratio (weight of water divided by weight of solid). Soil with a water content of 50% is one-third water (Figure 3.2).

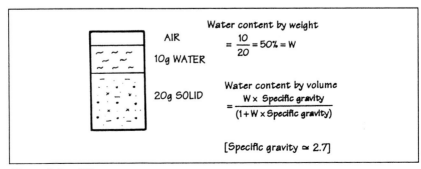

Figure 3.2   Water content.

Plasticity (or shrinkability)

Within the narrow scope of this book, the terms 'plasticity' and 'shrink-ability' are interchangeable; the latter is probably more commonly used. In soil mechanics generally, plasticity has a wider context.

Plasticity is an indirect measure of some of the ways in which a soil responds to water coming and going. Briefly, the more plastic, or shrinkable, the soil, the more it is likely to shrink and crack if water is removed by evaporation or by suction exerted by transpiring vegetation. The degree of plasticity is judged by measuring two values:

- the liquid limit, which is the water content above which the soil starts to flow as a liquid
- the plastic limit, which is the boundary, also defined by water content, between the mouldable (plastic) state and the friable (semi-solid) state.

These boundaries are not, like melting ice, instantly recognisable in nature; the changes from one state to another occur as imperceptible transitions. Fortunately, well-established laboratory tests are available that provide repeatable results, and these results produce a useful empirical classification. They do not, unfortunately, generate values that engineers can translate into exact predictions of behaviour, as we shall see.

Figure 3.3 shows the classification generally used. Following laboratory tests, the value for liquid limit is plotted on the horizontal scale and the value for liquid limit minus plastic limit (known as *plasticity index*) is plotted on the vertical scale. Most values for clay soil lie above and close to the diagonal 'A-line'. The higher up the A-line, the more plastic (or shrinkable) the soil is judged to be. Pure sands and gravels are non-plastic, and do not appear on the chart.

Particle size and water

Particle size and water interact as follows: water reduces the strength and stiffness of the soil, and particle size affects the movement of water through it. This interaction is weak in coarse soils. The strength of wet gravel is only slightly less than that of dry gravel, and the obstruction provided by gravel particles to the movement of water is minimal. Water moves freely through gravel. At the other end of the scale, the strength of clay can be reduced significantly if the water content increases. This can be deduced from its classification (Figure 3.3). Every clay has a plastic limit, below which it is too strong to mould by hand, and a liquid limit, above which it has virtually no strength. The rate of movement of water through clay is governed by permeability, and clay is typically a million times less permeable than sand. (Permeability is a sensitive property, and can vary by a factor of 1,000, even from one type of clay to another. Fissures and partings of silt or sand within clay effectively increase its value.)

When loading is applied to clay – for example, when a building is constructed upon it – water slowly drains from it, and it consolidates. This process is described in Figure 3.4 (*overleaf*).

Water and plasticity

The interaction between water and plasticity is negligible in the case of coarse soils, but clay is very much at its mercy.

Clay changes volume when its moisture content changes. Seasonal water content (moisture content) variations cause shrinkage and swelling to occur within 1m to 1.5m of ground level. Foundations constructed within the top metre or so experience shrinkage as a loss of support (subsidence), and swelling as a strong upward pressure (heave). They go through regular cycles of up-and-down movement, worse in some years than others, and occasional cosmetic damage may be the result. Trees considerably extend the zone of moisture change. As they grow, their root systems extend during the summer months, developing opportunistically

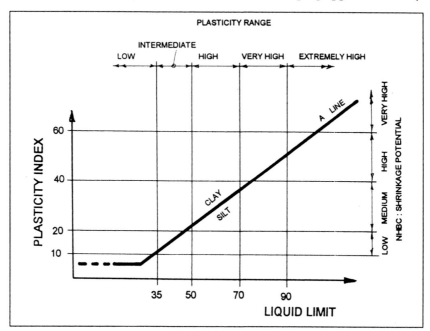

Figure 3.3   Plasticity.

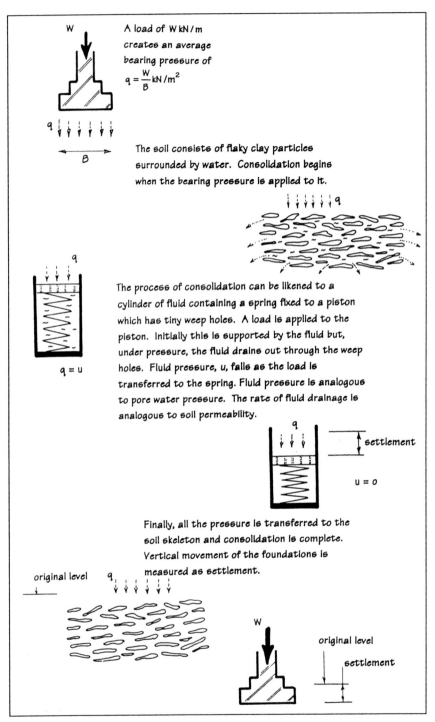

W

A load of W kN / m creates an average bearing pressure of

$$q = \frac{W}{B} kN / m^2$$

q

B

The soil consists of flaky clay particles surrounded by water. Consolidation begins when the bearing pressure is applied to it.

$\downarrow\downarrow\downarrow\downarrow\downarrow\downarrow$ q

q

q = u

The process of consolidation can be likened to a cylinder of fluid containing a spring fixed to a piston which has tiny weep holes. A load is applied to the piston. Initially this is supported by the fluid but, under pressure, the fluid drains out through the weep holes. Fluid pressure, u, falls as the load is transferred to the spring. Fluid pressure is analogous to pore water pressure. The rate of fluid drainage is analogous to soil permeability.

q

settlement

u = 0

Finally, all the pressure is transferred to the soil skeleton and consolidation is complete. Vertical movement of the foundations is measured as settlement.

original level

q

W

original level

settlement

Figure 3.4   Clay drainage analogy.

in soil with favourable amounts of water and oxygen. Roots are capable of extracting moisture at a distance of 750mm from immediately behind the growing tip. The behaviour of the soil/water mass under suction is similar to its behaviour under load (Figure 3.4), except that:

- It is driven not by external load but by suction; and
- There is no long-term equilibrium; water content reduction continues for as long as the vegetation needs water and the clay can provide it.

Figure 3.5 (*overleaf*) describes the process of suction and subsequent recovery.

It would be useful to be able to calculate volume changes from water content and plasticity. Unfortunately, we can attempt only rough approximations at best. The timescale is even less predictable. This makes life difficult for the building professional who is responsible for choosing an appropriate repair. It is all too easy for a person with less knowledge and responsibility, but plenty of hindsight, to find grounds for criticism at a later date. A link can certainly be made between plasticity and litigation!

Volume change in clay

Volume change in clay drives some of the more intractable foundation problems discussed in Part Two.

Volume change is driven by water content change. Figure 3.6 (*page 29*) shows the nearly linear relationship between water content and volume, from shrinkage limit to liquid limit. Shrinkage limit is the point below which any further reduction of water content has no effect on soil volume. Figure 3.6 also notes the point at which the voids between clay particles are completely filled with water. This is when the clay is said to be saturated. It is not, as the term might imply, the point at which no further intake of water is possible.

Figure 3.7 (*page 29*) shows the case of a clay that has already been dried to some extent by vegetation. The solid line shows how the current water content varies with depth. If the vegetation remains, the clay could, in theory, dry out to below its shrinkage limit. If the vegetation is removed, the clay's water content should rise to its level before the vegetation took hold. The latter is sometimes referred to as 'equilibrium water content'. In either case, we have a simple estimate for change in water content and, according to Figure 3.6, this should yield a simple estimate for soil volume change. In both cases, there are practical difficulties.

Usually, it is the equilibrium water content that most interests us. We can produce an equilibrium figure for the site under investigation by measuring the water content at regular depth intervals in a 'control borehole'

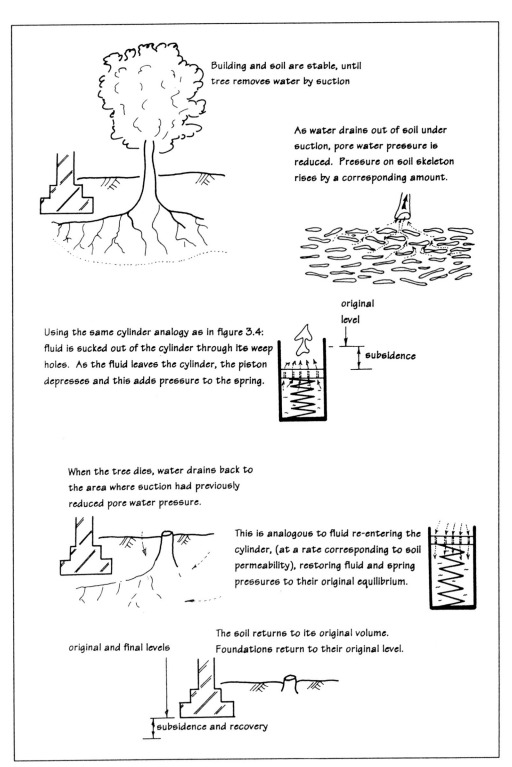

Building and soil are stable, until
tree removes water by suction

As water drains out of soil under
suction, pore water pressure is
reduced. Pressure on soil skeleton
rises by a corresponding amount.

original
level

subsidence

Using the same cylinder analogy as in figure 3.4:
fluid is sucked out of the cylinder through its weep
holes. As the fluid leaves the cylinder, the piston
depresses and this adds pressure to the spring.

When the tree dies, water drains back to
the area where suction had previously
reduced pore water pressure.

This is analogous to fluid re-entering the
cylinder, (at a rate corresponding to soil
permeability), restoring fluid and spring
pressures to their original equilibrium.

The soil returns to its original volume.
Foundations return to their original level.

original and final levels

subsidence and recovery

Figure 3.5   Clay suction analogy.

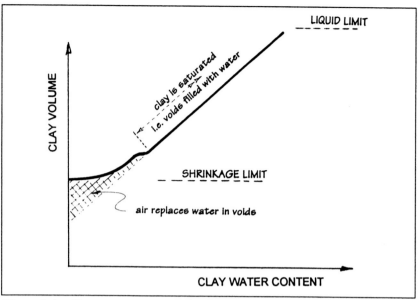

Figure 3.6   Clay water content and volume change.

as close as possible to where we want to calculate the volume change, but remote from the effects of any vegetation, if possible. The purpose of the control borehole is to represent conditions identical to those where we wish to calculate the volume change, apart from the vegetation – the

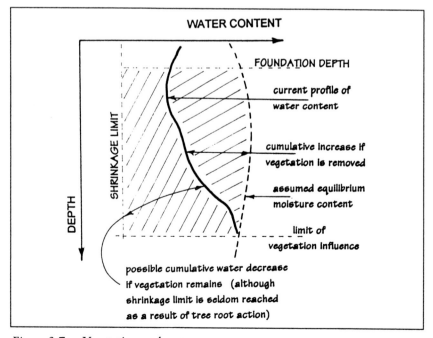

Figure 3.7   Vegetation and water content.

one factor that we might wish to change. If no such conditions can be located, a rough 'control' figure can be taken as 40% of liquid limit, but only for clay of very high or extremely high plasticity (Figure 3.3, *page 25*). In clay of lower plasticity, such as boulder clay, 2% above plastic limit is often taken as more representative. It cannot be overemphasized that these figures are far from precise.

Next, we need to convert water content change to soil volume change. We start this calculation with water content by *volume* (Figure 3.2, *page 23*), rather than by weight. This volume change is then divided by a number – the water shrinkage factor (the slope in Figure 3.6) – to get soil volume change. This number varies from clay to clay and from place to place. As a first guess, without specialist advice, a water shrinkage factor of between 3 and 4 may be used. (In most circumstances, a figure of 3 provides an overestimate, which may be a useful 'safe answer'.)

Few design calculations are less reliable than predicted clay volume change. There is no exact 'equilibrium' water content, because it is influenced by the geological history of the clay. No two samples of the same clay are identical. Every site has its own topography, storm water drainage characteristics, even microclimate. Laboratory testing, although sufficiently accurate for establishing basic properties like plasticity, has a 'working tolerance' sufficient to overshadow the true volume change in many cases.

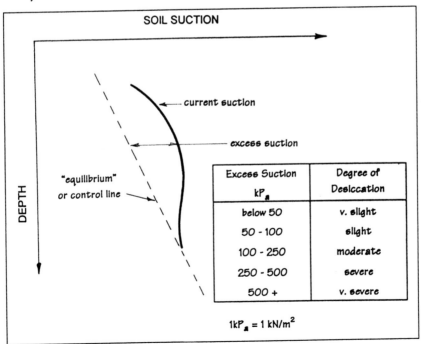

Excess Suction kP$_a$	Degree of Desiccation
below 50	v. slight
50 - 100	slight
100 - 250	moderate
250 - 500	severe
500 +	v. severe

$1kP_a = 1 \text{ kN/m}^2$

Figure 3.8   Clay suction and desiccation.

Suction testing (Figure 3.8) is a more direct measure of potential swelling. It suffers from the same site variations, but gives more reliable results than water content testing, so it is sometimes worth the extra cost. (For example, sampling needs to be more careful, and therefore it costs more.)

Mixed soils with high silt or sand contents, such as boulder clays, give less reliable results than even the 'heavier' overconsolidated clays. This applies to all test methods, including suction.

Cohesion and friction

The strength of clay derives from its cohesion – the ability of its individual particles to stick together despite forces of disruption. In contrast, the strength of granular soils (sand and gravel) derives from internal friction – the resistance of their individual particles to forces causing them to slide across each other. Tables 3.2 and 3.3 give common descriptive terms to measurements of cohesion and internal friction.

Table 3.2   Strength of cohesive soil

	Cohesion kN/m^2
Very soft	Less than 20
Soft	20 – 40
Firm	40 – 75
Stiff	75 – 150
Very stiff	150 – 300
Hard	More than 300

Table 3.3   Strength of granular soil

Term	Angle of internal friction (degrees)
Very loose	Less than 29
Loose	29 – 30
Medium dense	30 – 36
Dense	36 – 41
Very dense	More than 41

Soil strength (cohesion or friction) determines its ability to support buildings and to remain stable on slopes, as discussed later in this chapter.

Table 3.4 summarizes some of the differences in properties between fine and coarse soils.

Table 3.4   Properties of fine and coarse soils

Fine soil	Coarse soil
Strength depends mainly on cohesion (particles binding together)	Strength depends mainly on friction (particles resisting sliding across each other)
Strength considerably reduced by increase in water content	Strength slightly reduced by increase in water content
Change in water content causes change in volume (according to plasticity)	Change in water content does not affect volume
Very low permeability	High permeability
Consolidates slowly	Consolidates quickly

Although silt is classified as a fine soil, it acts in many ways as a transition between fine and coarse. Coarse silt derives its strength mainly from internal friction; fine silt is cohesive.

Mixtures and layers

Even in such a brief introduction, we must admit one or two complications. Most soils are mixtures (Figure 3.1, *page 23*). Sandy clay, for example, is a term describing a soil that is predominantly clay but also contains a significant proportion of sand. The sand content would increase permeability and reduce plasticity. Clayey sand would also have intermediate properties, but they would be closer to sand. A soil with 35% fine particles and 65% coarse would have one foot in both camps. Its strength would derive from both cohesion and friction, and would be affected to a moderate extent by water content. The soil would probably have some plasticity, but it would plot towards the lower end of the 'A line' (Figure 3.3, *page 25*), although only testing would tell.

Mixtures of silt and fine sand (Figure 3.1) are often close to the 65–35 division. These soils have a few properties peculiar to themselves. Their unfortunate combination of particle size and permeability makes them more vulnerable than other natural soils to:

- erosion by running water
- sudden volume loss if the water table is lowered by pumping

- volume loss caused by vibration
- heave caused by frost.

On many sites, the layer of soil immediately beneath the building is thick enough to be the only influence on foundation behaviour. Where two or more strata of differing properties are within 'influencing distance' of the foundations, behaviour may be difficult to predict. A sand layer between clay strata might dry the clay by draining it, or it might, if water bearing, increase the clay's water content. In the latter case, it would replenish water removed by vegetation.

Soil conditions can also vary over short horizontal distances. Alternations between strong and weak (Figure 3.9) can be a greater problem than soil that is consistently weak. The same is true of alternations between plastic and non-plastic. In both cases, these abrupt changes exaggerate differential movement; it is differential movement, rather than average or maximum movement, that causes structural damage.

Peat and organic soils

Peat deserves special mention because it is a serious hazard. It consists of plant remains, sometimes mixed with silt, clay and other inorganic particles. There are many types of peat. They vary in degree of plant decomposition (humification) and in the amount and type of inorganic content. As long as the organic content exceeds 50%, all types of peats can be broadly classified as fine soils with very high plasticity (highly shrinkable) and a very high water content, usually above 100%. The high water content makes them weak compared with most inorganic soils. To add to

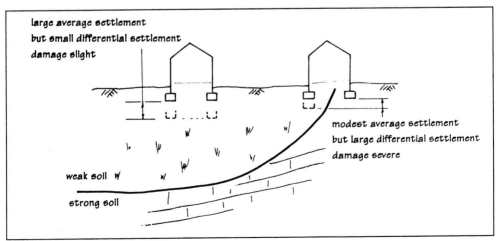

Figure 3.9   Variable soil.

these disconcerting properties, peat is extremely variable, both in its stratum thickness and its material properties. Foundations built on peat may settle abnormally and excessively, and for a prolonged period. They will certainly be prone to subsidence arising from shrinking, if moisture is removed by vegetation or drainage. Moisture removal can promote a permanent change in behaviour, as the peat becomes oxidized when the water recedes. This reduces its capacity for re-absorbing water, and it may also reduce its strength.

When a peat layer is first exposed by excavation, a musty odour is often released, sometimes (but not always) accompanied by a 'bad egg' smell from hydrogen sulphide. The odour may soon fade. The excavated hole will often quickly fill with water. Peat is also recognizable by its dark colour and light weight. Its plant origins may be obvious from its fibrous structure, but not if decomposition is advanced. In the latter case, the plant remains may be amorphous, but can be identified as organic by thinly smearing a small sample across a clean surface.

Some soils have a low organic content; their behaviour as an inorganic soil, influenced by particle size and water, will be only slightly modified by the organic content. The organic content may increase the soil's plasticity (shrinkability) or reduce strength, or both. Organic soil is often darker and smellier than inorganic soil. The darkness is sometimes patchy or spotty.

Made ground

Made ground is soil deposited by man. It includes natural inorganic soil, simply redeposited, sometimes well compacted, sometimes tipped; and it includes all the waste products of civilization. There is an enormous range of materials waiting to be found. There is space here to mention only a few of the notable characteristics of made ground.

Most made ground has had very little time to consolidate compared to natural soil. It may still be consolidating under self-weight. It may be weak, and so settle more than is usual under the weight of a new building. It may continue to settle for longer than natural soil, particularly if it is still undergoing chemical reactions. And it may be more than usually vulnerable to subsidence caused by changes in the immediate environment. If it contains collapsible material (such as empty drums or old cars) or degradable material (such as domestic refuse), its settlement – especially differential settlement – will be greatly increased. Certain types of made ground, if placed dry and with little or no compaction, can suffer large and sudden volume reduction if water enters it at a later date (Chapter 9).

Contamination and pollution

Contamination was not taken very seriously until the 1970s when, close to the town of Niagara Falls, New York, an abandoned canal filled with chemical waste leaked its contents into water courses and the basements of nearby houses, causing ill health to residents. It soon became apparent that certain industries (such as gasworks, scrapyards, sewage works, tanneries and waste disposal) had left a legacy of pollution on the land they had occupied; this needed to be cleaned or rendered harmless to the next users, as well as to any neighbours downstream or otherwise within range of the contaminants. Methods of investigation and treatment began to be developed. A wider range of contaminants were recognized.

The learning curve steepened during the late 1980s and early 1990s. Recognition and testing are now carried out more consistently. Knowledge of remediation, its needs and methods, will continue to improve.

Some buildings originally constructed on clean soil are now found to be on contaminated soil as a result of contaminants migrating from other areas. Some existing building stock is on contaminated ground whose unwelcome properties were underestimated, and very likely unrecorded, at the time of construction. Contamination can, therefore, come as a nasty surprise during repair work. The expected standard of cleaning up will be very much higher than it was at the time of construction, if indeed any cleaning up at all was expected then. Both soil and ground water may be affected by contaminants; hazardous gas may be present. Designers have a duty to protect those carrying out repairs and anyone else who might be exposed to harm. If a possible contamination hazard is perceived, specialist advice must be obtained (Table 6.3, *pages 62–63*).

Not all contaminants are man-made. Radon is a colourless, odourless gas that is present in minute harmless amounts in most buildings, but occurs in sufficient quantity to be a recognized hazard in certain locations, such as southwest England. While this is not a structural problem, it is often convenient to tackle it at the same time as structural repairs.

Diversity of soil

To the foundation engineer, soil consists of any material close enough to the surface to affect the behaviour of a man-made structure. In the UK, such material includes strong rocks, like granite, weak rocks, like chalk, and the products of their erosion: gravel, sand, silt and clay. Geological faulting and climate changes have created abrupt variations in land form, soil type and soil behaviour. Glaciation mixed different materials and transported them over long distances, sometimes compacting them into strong deposits, sometimes dumping them in weak, marginally stable slopes. Marine and alluvial action continues the process of erosion and deposition to this day.

A glance at any geological drift map, or soil survey map, confirms the complexity of the soil closest to the surface. It is often difficult to be sure, by reference to maps alone, whether a site is on one or more of several soils with radically different properties. Strata thicknesses are even harder to judge. Even borehole drilling does not guarantee to find everything of interest on site.

The effect of soil on building

The building needs to be:

- supported safely by the soil
- protected from any independent movement of the soil.

Support relies on the strength of the soil (friction or cohesion, or both if both are present) and the geometry and depth of the foundations. Since the foundations themselves form part of the equation, it is not possible to give precise formulae for safe bearing capacities (also known as allowable bearing pressures), but Table 3.5 gives a first guess. Only natural soil is represented in this table. Well-compacted made ground can be as strong as its natural counterpart, but few foundation designers nowadays would trust it without specifying some improvement and testing it for contamination.

The values in Table 3.5 are rough, because soil and rock vary in quality; other factors, unique to each site, will also have some influence. One such factor is the depth of foundations below ground level – a factor with little influence in the case of granite, but a dominant factor in the case of clay. The mechanism of failure in clay has practical implications of enormous importance to underpinning (discussed in Chapter 17). Figure 3.10 (*page 38*) illustrates this mechanism.

Slopes

Assessing the stability of slopes requires specialist knowledge, and most foundation engineers consult a geotechnical engineer when they perceive a problem. As noted at the beginning of this chapter, soil particle size, water and plasticity are major influences on soil behaviour. The geotechnical engineer is also concerned with soil structure and geology, whose evaluation requires a higher order of experience and judgement. Figure

Table 3.5   Rough guide to allowable bearing pressure

Soil or rock	Typical allowable bearing pressure kN/m^2
Granite *(strong)*	5×10^3 to 10×10^3
Sandstone *(moderately strong)*	1×10^3 to 4×10^3
Limestone *(moderately strong)*	0.5×10^3 to 6×10^3
Schists and slates *(moderately weak)*	2 to 3×10^3
Shales *(moderately weak)*	1 to 2×10^3
Mudstone	50 to 400 depending on grade
Chalk	50 to 300 depending on grade
Very dense sand or gravel	500
Dense sand or gravel	300
Medium dense sand or gravel	150
Loose sand or gravel	50
Dense silt	120
Medium dense silt	80
Loose silt	25
Hard clay	500
Very stiff clay	300
Stiff clay	200
Firm clay	100
Soft clay	50
Very soft clay	25

Notes:
1. The bearing pressure of thinly bedded sandstones and limestones, and of shattered rocks and of weathered chalk are unpredictable, and values must be based on thorough investigation.
2. The bearing pressures of silts, sands and gravels are reduced (by roughly 50% as a first guess) if ground water is, or can rise to, as close to the underside of the foundations as their width.
3. The values for clay, silt, sand and gravel are increased if the depth of foundations exceeds 1m (see Figure 2.10, *page 15*).

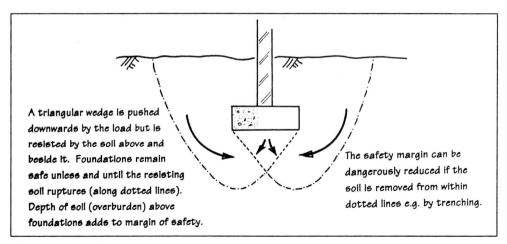

A triangular wedge is pushed downwards by the load but is resisted by the soil above and beside it. Foundations remain safe unless and until the resisting soil ruptures (along dotted lines). Depth of soil (overburden) above foundations adds to margin of safety.

The safety margin can be dangerously reduced if the soil is removed from within dotted lines e.g. by trenching.

Figure 3.10   Failure in clay.

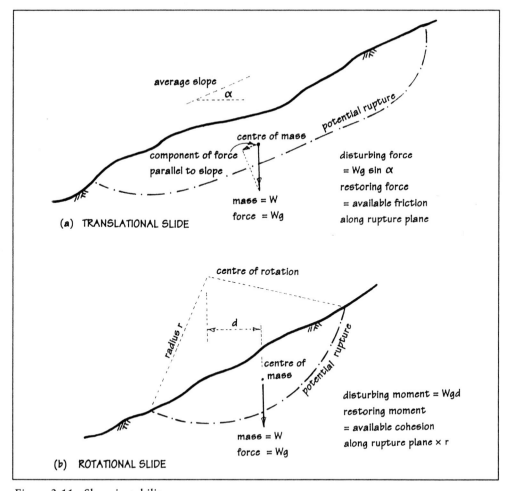

average slope
α

centre of mass

potential rupture

component of force parallel to slope

disturbing force
= Wg sin α
restoring force
= available friction
along rupture plane

mass = W
force = Wg

(a) TRANSLATIONAL SLIDE

centre of rotation

radius r

d

centre of mass

potential rupture

disturbing moment = Wgd
restoring moment
= available cohesion
along rupture plane × r

mass = W
force = Wg

(b) ROTATIONAL SLIDE

Figure 3.11   Slope instability.

3.11 illustrates two common types of instability. Most incidents are more complex than these simple models.

As long as the forces tending to disturb a slope are met with adequate resistance, generated by the soil's strength, the slope remains safe. The extent to which resistance exceeds disturbance creates the margin of safety. For many natural slopes, this margin is sufficient but not generous, and the soil can be accidentally destabilized. Figure 3.12 shows that adding weight uphill of the centre of rotation (A) will reduce the margin of safety by encouraging the 'pendulum of disturbed soil' to swing. Cutting trenches, for example for drainage, parallel to contours (B), may allow the potential rupture to bypass some of the soil resistance, perhaps only by a short length, but a short length may be enough to remove the margin of safety. Removal of vegetation (C2) from clay soil would lead to an increase in water content (Figure 3.5, *page 28*), with a consequent reduction in cohesive strength, and therefore a reduction in the margin of safety. Because of the clay's very low permeability, it would take a long time for this to take full effect. An unattended water leak (C2) can weaken soil of any type.

Stability can be improved by adding weight downhill of the centre of rotation (D), or by providing a strong barrier (E) across potential slip planes. These barriers might consist of ground anchors, a massive wall or sheet piles anchored safely on the other side of the potential slip plane. Drainage, particularly deep drainage, should improve soil strength by reducing pore water pressure. It needs careful design and maintenance if it is to be permanently beneficial.

A slope stability analysis needs to discover the unique line along which slippage is most likely to occur. This is a difficult task, but necessary, because any other line assumed in the assessment will give a falsely optimistic

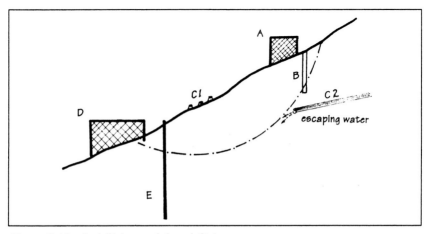

Figure 3.12   Stabilizing and destabilizing agents.

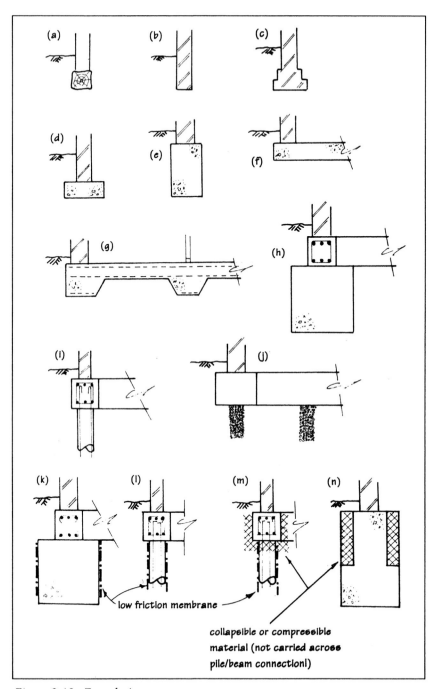

low friction membrane

collapsible or compressible
material (not carried across
pile/beam connectionl)

Figure 3.13   Foundation types.

answer. Discovery is by trial and error. Computers reduce the drudgery of calculation, but the geotechnical engineer still has to take into account many factors such as local weakness, fissures, and existing shear planes (which are often the ghosts of previous slippages). It is very difficult to establish the likely strength of the soil along each of its many potential failure paths – hence the need for specialist advice.

Foundations

Figure 3.13 shows the main types of foundation. The most common is the strip footing, where the load-bearing walls make line contact with the soil through a sole plate (a), plain (b), or corbelled (c) brick footings, or concrete strip (d). From the 1960s onwards, trench fill (e) became a popular version of the continuous strip, because it offered a simple means of bypassing shallow hazards without the labour-intensive task of laying bricks below ground level.

Concrete rafts have been used for soft ground, with varying success. Before 1950, most rafts were thin (150mm or so) and unreinforced (f). Later, reinforcement was used more often, and almost invariably after 1970. Various types of stiffening ribs were introduced to lessen differential settlement (g).

Deep concrete pads (h) (most within 2 to 5 metres of ground level) were used to bypass hazards of moderate depth; the building loads were transferred to them through reinforced concrete beams (ground beams).

Piling offered an alternative to pads where loads were heavy or hazards were deep (i). There are many types of pile, some transferring their loads at their base only (end bearing piles), some transferring them via shaft as well as base (friction piles). The most common material used has been concrete. Steel piles are the second favourite material and, for buildings constructed in the first half of the twentieth century, timber piles have also been used.

Since the 1970s, ground treatment has become more popular. Its purpose is to raise the strength (and therefore the allowable bearing pressure) of weak soil to a modest level, and to reduce its variability – thereby reducing the differential settlement of anything built on it. The methods include pre-loading, dynamic compaction and vibro methods, which introduce stone or concrete columns into the ground at close centres (j). Foundations can then be a continuous strip, usually reinforced, or a reinforced concrete raft.

Earlier foundations were expected to support only the building. The need, sometimes, to protect the building from independent ground movement (subsidence, landslip or heave) was not perceived. Foundations should nowadays be designed for this need, whenever it arises. A raft foundation may be designed to tolerate subsidence, although bypassing it with deep foundations, such as (h) and (i), is usually more acceptable. (For deep mining subsidence, raft foundations are more appropriate.) Subsiding soil may drag on the sides of foundations that penetrate it, in which case the foundations can either be wrapped in a low-friction membrane, such as in (k) and (l), or be designed to support the additional load (downdrag).

Low-friction membranes cannot eliminate downdrag altogether, so a calculation is always needed. It is seldom possible to accommodate landslip. Attempts to do so, for example by using rafts (g), have often been disappointing. In the case of heave, collapsible or resilient 'anti-heave precautions' can be fixed wherever the clay might expand vertically or laterally against the foundations, as in (m) and (n), to eliminate or reduce its pressure. If no more than a reduction of its pressure is achieved, the foundations must be designed to withstand the residual force.

This brief review has not covered all the types of substructure in existence. In the nineteenth and early twentieth centuries, various theories for strengthening the footing or spreading the load were put into practice; two of the most popular solutions were grillages, often of timber, and arches, usually of brickwork. Expert advice should be sought if anything is not recognized when it is discovered during investigation or repair work.

Chapter Four

Uncertainty and performance

Calculations for forces, stresses and strains have a spurious precision. Every structural engineer knows that they embody a great deal of uncertainty and are not intended to predict performance. This chapter describes how uncertainty is managed in the design of new buildings, and explains how the frame of reference changes in the case of existing buildings and their repair. In most cases, existing buildings enjoy a reduced uncertainty, and this encourages economies in repair. This is not inevitable. Small details can impair performance.

Structural modelling

Structural analysis always has to be based on a model. This may be a simplified outline of the structure (such as that produced in Figure 2.4, *page 10*), with sufficient information on loads and dimensions to enable the first steps in calculation to be made. Models suffer from the following inaccuracies:

- Accidental support is not counted.
- The benefit of 'non-structural' elements is not counted.
- Secondary load paths and forces are difficult to model.
- Complex details usually have to be simplified.

The net result of these inaccuracies is usually, but not always, an underestimation of performance.

Soil–structure interaction

Simple models also ignore the soil's influence. They make an unstated assumption that the loads are accepted passively by the soil, which develops a linear resistance of appropriate value under every wall, as well as a local point resistance to match every point load from piers or columns. In practice, this can only happen if the soil is infinitely stiff compared to the building. A wooden shed on granite approaches this condition. Figure 4.1a illustrates the case. 'Load from soil' is a mirror image of 'load from building.' (To keep the units compatible, load from soil is assumed to be soil bearing pressure times foundation width; thus both up and down loads can be expressed in kN/m.)

The opposite extreme – soil infinitely flexible compared to the building it supports – calls for a considerable redistribution of soil bearing pressure, with a linear variation between maximum and minimum (Figure 4.1b). This condition might be pictured, if disbelief can be suspended for a moment, as a skyscraper on mud.

Real cases lie between the two extremes (Figure 4.1c). Every case is unique. The actual bearing pressure distribution is predetermined by the combination of building and soil properties peculiar to each site; its pattern can be significantly changed only by major structural repair or alteration.

It is not every day that soil–structure interaction forms part of the calculations for repairs. The subject has been introduced mainly to draw attention to its effect on building behaviour, which might otherwise, in some circumstances, seem inexplicable. A greater understanding is essential only to specialists needing to make a detailed appraisal of a damaged building or design a sensitive repair. As an introduction for those who may wish to take the subject a little further, Figure 4.2 provides a qualitative description of how each building must find a way to redistribute

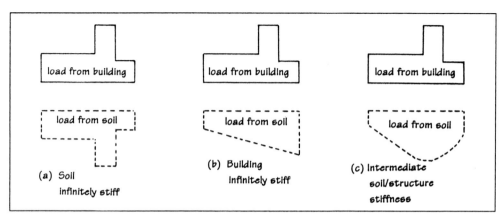

Figure 4.1   Soil–structure stiffness.

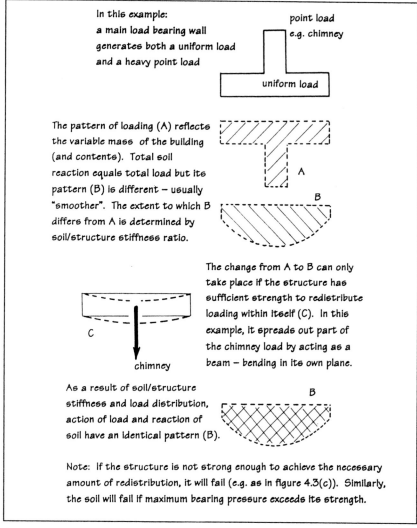

In this example:
a main load bearing wall
generates both a uniform load
and a heavy point load

point load
e.g. chimney

uniform load

The pattern of loading (A) reflects
the variable mass of the building
(and contents). Total soil
reaction equals total load but its
pattern (B) is different – usually
"smoother". The extent to which B
differs from A is determined by
soil/structure stiffness ratio.

A

B

The change from A to B can only
take place if the structure has
sufficient strength to redistribute
loading within itself (C). In this
example, it spreads out part of
the chimney load by acting as a
beam – bending in its own plane.

C

chimney

As a result of soil/structure
stiffness and load distribution,
action of load and reaction of
soil have an identical pattern (B).

B

Note: If the structure is not strong enough to achieve the necessary
amount of redistribution, it will fail (e.g. as in figure 4.3(c)). Similarly,
the soil will fail if maximum bearing pressure exceeds its strength.

Figure 4.2   Bearing pressure and load redistribution.

forces through itself, in order to match the soil bearing pressure de-
manded by the soil–structure stiffness ratio. For new buildings, this redistribution is usually entrusted entirely to the foundations. In old buildings,
with shallow and relatively weak foundations, the stiffness and strength
of the superstructure was, in the past, inevitably called into play. (Such
redistribution should now be complete, unless there have been recent
changes to the building or its environment, or to soil volume).

Figure 4.3 (*overleaf*) crudely compares the behaviour of strong and
weak structures coming to terms with low soil–structure stiffness ratios.
It is no coincidence that many examples of both tilts and breaks can be
seen in Fenland, where there are large areas of marine alluvium and peat

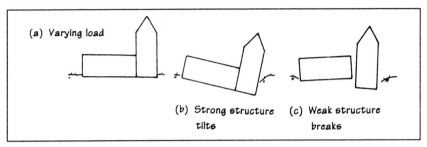

Figure 4.3   Soil–structure stiffness.

and, hence, often low soil–structure stiffness. (A tilt is not inevitably caused by eccentricity of the building's loading. Less often, but also less recognisably, it can be caused by variations within the soil.)

Modelling stiffness

Redistribution is generated not only by soil–structure interaction. Stiffness ratios within the superstructure itself will also affect the routes taken by the loading as it passes downwards, creating forces, stresses and strains within the individual members. This effect has not yet been addressed in this book. Table 2.1 (*page 6*) implies that stiffness affects only strains (distortions). This is an oversimplification. Although the resulting inaccuracy is too small to matter for many simple structures and simple elements, a knowledge of structural engineering is needed to judge whether the analysis should be more sophisticated. The level of structural engineering required is, unfortunately, beyond the scope of this book. (It is covered in *Understanding Structural Analysis* by David Brohn; see *Further reading.*) For the present purpose, suffice it to say that loading tends to be attracted towards stiffer parts of the structure. To give the briefest of examples: the large chimney indicated in Figure 4.2 would probably attract most of the wind loading on that part of the building. This would be to the advantage of the building as a whole, provided that the chimney has the strength to cope with the stresses caused by the acquired loading.

Uncertainty

Material strength depends on many factors. With brickwork, for example, one factor is the quality of the mortar in its joints. The mortar quality, in turn, depends on the grading of the sand and the shape of individual

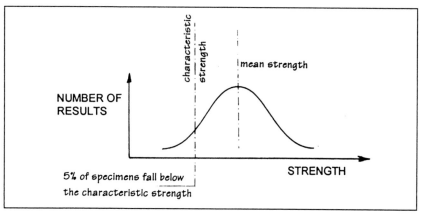

Figure 4.4   Typical distribution curve.

particles, any extraneous fine matter, proportions of cement or lime, water content, any additives, time between mixing and using, speed of construction, weather conditions, suction of the brick, and skill in applying mortar to brick and brick to mortar.

These and many other uncertainties about performance are bundled into a statistical treatment, so that the structural engineer does not have to endure a prolonged meditation before starting every design. The basis of this statistical treatment is the distribution curve.

If a large number of material samples are tested to failure, a distribution curve can be formed from the results to describe their variation in strength (Figure 4.4). Designers would like to be able to assume a strength that will be absolutely safe, but Figure 4.4 shows this to be impossible. The flattening of the curve at either end is an indication that, although extreme values are few, it is not realistic to state either a minimum or a maximum. It is possible to define only values that are exceeded by a given percentage of results. In Figure 4.4, the 'characteristic strength' is the term applied to the value that is exceeded by 95% of test results. A member whose stress equals the characteristic strength has a one in twenty chance of failing (assuming, to keep the argument simple, that no unwelcome influences, such as slenderness, apply).

The curve in Figure 4.4 is typical of most building materials. Obviously, the numerical values vary from one material to another, and some curves have a more compact shape than others. Steel is man-made and aimed to fit the established curve, which is therefore compact, or thin and tall. Timber is organic. We have to accept what comes, and its curve is flatter; in other words, its ratio of mean strength to characteristic strength is greater than is the case with steel.

Loading also follows a distribution curve. Again, it is not possible to define, for example, a wind load that will never be exceeded. It would be nonsense to impose the highest floor loading that is physically achievable, but we can produce characteristic values. The aim must then be to ensure that, in every member, the stress caused by characteristic loading is exceeded by the characteristic strength of the material. This policy is illustrated in Figure 4.5. The characteristic value for the 'loading' curve is the one exceeded by only 5%. If the two characteristic values were equal, the statistical probability of failure (by fracture, buckling or whatever) would be 5% times 5%, equal to 1 in 400. Since the successful performance of a building relies on several different distribution curves, better odds than 1 in 400 are preferred. Figure 4.5 shows the characteristic values separated by a gap representing additional safety (over and above the use of the characteristic values themselves) built into the design. The gap is forged by 'partial safety factors'. Every load and every material is given a partial safety factor whose value is set in accordance with convention.

Despite characteristic values and partial safety factors, there will always be a tiny blurred overlap (the shaded area in Figure 4.5) of the extreme 'tails' of the strength and loading distribution curves. These tails represent residual risk. Absolute safety cannot be guaranteed; we can only aim at a risk of failure that is acceptably low.

Modelling repairs with uncertainty

Structural repair follows the same principles as design – equilibrium must be assured, based on modelling and taking account of uncertainty – but the viewpoint is different. The structure's performance can be observed. Any structural damage of course represents a vanished margin of safety,

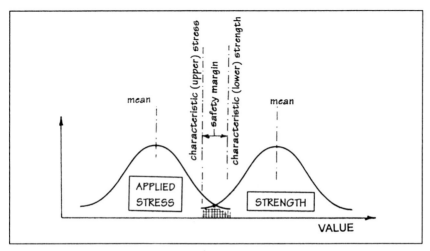

Figure 4.5   Safety margins and risk.

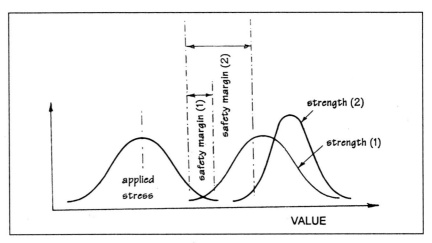

Figure 4.6   Uncertainty reviewed.

but this is usually local and instructive. The model is based on observation of performance, including locally unacceptable performance, and it can be more sophisticated than the equivalent model for a proposed building. Accidental supports, secondary forces, 'non-structural' elements and the effect of complex details can all be counted. Thus uncertainty can be reviewed.

Whereas a new member must be designed according to its characteristic strength, an existing one can be credited with its actual strength, at least to the extent that this can be judged by inspection, performance or testing. The idea is illustrated, in abstract form, in Figure 4.6, which translates the earlier strength distribution curves from new building to repair. The curve marked 'strength (1)' represents the ability of the building as designed to withstand the loading. The curve marked 'strength (2)' represents the current assessment of the same building's ability to withstand loading, now that we can see the actual quality of its materials (masonry, pointing, timber, and so on) and how it has performed. We can see how it has settled, cracked, distorted and perhaps deteriorated. The 'strength (2)' curve, in this example, is further to the right and more narrow than 'strength (1)'. This reflects better performance than assumed in the original design and reduced uncertainty, which is the typical (but by no means inevitable) verdict of appraisal.

To what extent can we take into account improved modelling and reduced uncertainty? Caution should continue to prevail. If we wish to use a model that reflects actual conditions, we must be all the more careful to think through each detail (Figure 4.7, *overleaf*) and never abandon uncertainty.

It is rarely logical to reassess loading. The fact that a slender exposed wall has inexplicably withstood one hundred years of wind may justify

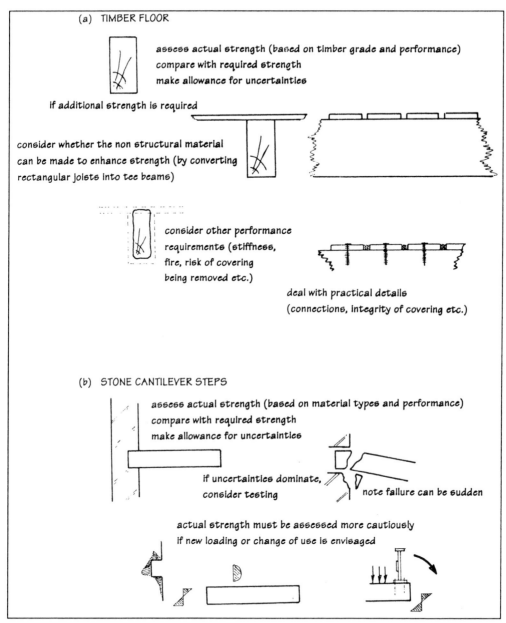

(a) TIMBER FLOOR

assess actual strength (based on timber grade and performance)
compare with required strength
make allowance for uncertainties

if additional strength is required

consider whether the non structural material
can be made to enhance strength (by converting
rectangular joists into tee beams)

consider other performance
requirements (stiffness,
fire, risk of covering
being removed etc.)

deal with practical details
(connections, integrity of covering etc.)

(b) STONE CANTILEVER STEPS

assess actual strength (based on material types and performance)
compare with required strength
make allowance for uncertainties

if uncertainties dominate,
consider testing

note failure can be sudden

actual strength must be assessed more cautiously
if new loading or change of use is envisaged

Figure 4.7   Design based on assessment.

re-examining its material strength, but it can never justify assuming a
wind load below what would be calculated from the British Standard.

Furthermore, if assessment is based at least partly on performance, the
building should of course be old enough to have experienced a range of
imposed loading. Even then, it should not be assumed that imposed loads
have ever reached their characteristic values, unless evidence is at hand.

Much of the time, imposed loading is trivial. In case of doubt, especially when it might be perilous to reduce uncertainty without good reason (for instance, because the building's use is to be changed, introducing higher loading), it is better to carry out tests (Figure 4.7b).

There are times when modelling the existing structure puts a brake on proposals, rather than supporting them. The obvious example is the occasion when actual material strength is below characteristic strength. Less obvious is the building whose behaviour has been enhanced by its so-called non-structural parts. Timber frame buildings often derive some of their strength and much of their stiffness and robustness from their infill panels, whose casual removal as part of a change or refurbishment could invite unexpected distortion or damage.

Finally, to emphasize the importance of observation, Figure 4.8 (*over-leaf*) shows how performance of the very simplest of structures can change over time, as a result of small changes of detail coupled with nature taking its course – something to watch out for when carrying out diagnosis (Chapter 7) or appraisal (Chapter 24).

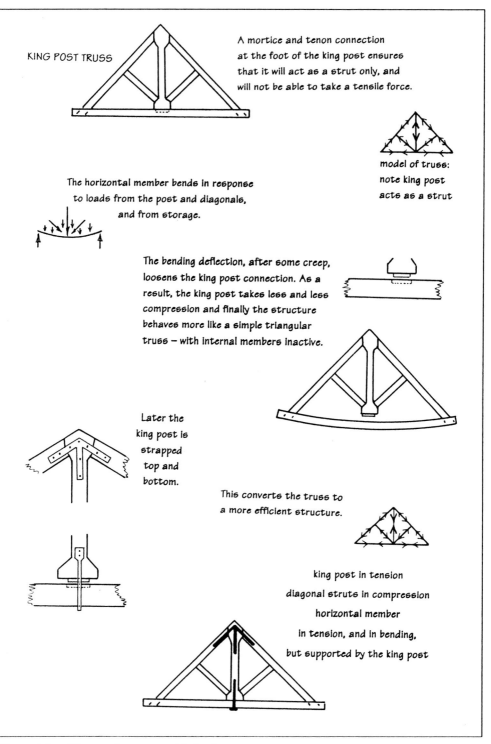

KING POST TRUSS

A mortice and tenon connection at the foot of the king post ensures that it will act as a strut only, and will not be able to take a tensile force.

model of truss: note king post acts as a strut

The horizontal member bends in response to loads from the post and diagonals, and from storage.

The bending deflection, after some creep, loosens the king post connection. As a result, the king post takes less and less compression and finally the structure behaves more like a simple triangular truss — with internal members inactive.

Later the king post is strapped top and bottom.

This converts the truss to a more efficient structure.

king post in tension

diagonal struts in compression

horizontal member in tension, and in bending, but supported by the king post

Figure 4.8   Changes in structural behaviour.

Chapter Five

Conservation

*Buildings of special architectural or historical interest, including virtu-
ally all buildings dating from 1700 or earlier and a few recent buildings
of outstanding value, may not be altered without permission. If permission is given, it will normally be subject to the rules of conservation, which
this chapter briefly introduces.*

Conservation is driven by its tenets, described as follows in clause 1.1
of BS 7913:1998 (*Guide to the principles of the conservation of historic
buildings*).

> The immediate and obvious objective of building conservation is to
> secure the preservation of the nation's stock of buildings, and in par-
> ticular its historic buildings and fine architecture, in the long term
> interest of society.
>
> The underlying objectives are cultural, economic and environ-
> mental.
>
> Attempts to separate out the objectives of conservation into distinct
> components, however, are usually unsatisfactory, since the issues are
> complex and often interlocked.

Table 5.1 summarizes the main principles of conservation. The rigour
with which these are applied varies according to individual buildings and
individual humans. At first sight, the principles seem to be at odds with
almost any structural repair, especially with high intervention work such
as refurbishment (Chapter 24). Many past refurbishments would be
denied listed building consent today, if the advice in BS 7913:1998 and
other current guidance were to be followed. But, just as good design has
to reconcile performance with cost and safety, so it can – indeed must –
reconcile performance with conservation. As with cost and safety, the
needs of conservation vary from one property to another. These needs are,
in some cases, necessarily restrictive, because the least intervention would

devalue the building. In other cases, a pedantic interpretation of the principles would obstruct urgent work.

Table 5.1   Conservation: outline principles

Ensure that all work has local planning authority consent.
Do nothing unnecessary.
Avoid damage to cherished details.
Match materials like for like.
Avoid modern materials if possible.
If modern materials have to be used, avoid clumsy disguise.
Use reversible methods of repair if possible.

Conservation encouragers

Without help, conservation would be a hobby, and redevelopment would eventually destroy most of the historic environment. Help is organized as follows:

- There is a system for identifying buildings and other features of historical or aesthetic interest (Table 5.2).
- Rules are provided for the conservation of these buildings.
- There are penalties for bad practice and grants for good practice.

Table 5.2   Listing and scheduling

Ancient monuments	Scheduled monument consent is required for maintenance work or alteration.
Historic buildings	Local planning authority consent is required for any work that might alter the character of the building.
Conservation areas	All demolition work requires consent. Other work might be affected by planning policy.
Archaeological sites	Local authority permission is required for any work that might disturb anything of archaeological interest.

There are many publications that describe how the protective system works (see *Further reading*). In addition, the local planning department or conservation officer will normally be able to give practical advice, if asked in good time.

If unauthorized work is carried out on anything scheduled, listed or designated, the offender may be prosecuted. There are also powers of deterrence that can be applied to non-listed buildings. Any individual or local authority can apply to the government (Department of the Environment, Transport and the Regions) to *spot list* a property thought to be in danger. If the danger is imminent, a quicker (and therefore probably more successful) form of protection would be a *building preservation notice*, a procedure instituted by the local planning authority, which has immediate effect.

The local planning authority can demand the repair of a neglected listed building by serving a *repairs notice*, specifying the works considered necessary and proper for its preservation. If the notice is not complied with, the authority may start compulsory purchase proceedings. For emergency repairs (for example, keeping the building stable or watertight) the local authority may empower itself to do the work and recover the cost subsequently from the owner.

Management of listed buildings

Some buildings, because of their function, undergo frequent minor changes. Each change requires listed building consent. Buildings needing frequent minor changes include large offices, industrial buildings, precincts, department stores, institutional buildings and housing developments. English Heritage recommends that owners of such buildings develop guidelines, in collaboration with local planning authorities, for consistent management. An agreed set of guidelines is almost guaranteed to improve the standard of conservation and save time and trouble on every building contract.

Chapter Six

Health and safety

The accident rate is higher in construction than in most industries, and it is especially high in maintenance, demolition and repair work. This chapter briefly describes the most important legislation and discusses how designers can make their contribution to health and safety in the repair and subsequent use of buildings.

Most fatal accidents in construction are caused by people falling, being hit by falling objects or being struck by mobile plant (Table 6.1). Young people, people new to the industry and people newly arrived on site and unfamiliar with its layout and routines are the most vulnerable.

Table 6.1   Causes of fatal accidents

Falls	Off roofs and through roofs (fragile lights)
	From scaffolds and working platforms
	From ladders
	Into excavations
	Into water
Falling objects	Collapse of structures
	Collapse of excavations
	Insecure loads, equipment and plant
Mobile plant	Struck by moving vehicles
	Struck by craned loads
	Falls from vehicles
	Vehicles overturning
Other causes	Electrocution
	Asphyxiation
	Fires and explosions

Through legislation, education and prosecution, efforts have been made to change the culture of health and safety from that of a secondary discipline, whose purpose was often defeated by the macho disposition of many senior members of the industry, to an essential part of both management practice and craft knowledge. Three changes of emphasis typified this culture change:

- Purposes and principles of health and safety management have replaced simple prescriptive rules.
- Responsibility for health and safety practice has been widened and better defined.
- Greater emphasis has been placed on long-term health.

The major acts and regulations that guided and reflected this change are listed in Table 6.2.

Table 6.2   Main legislation 1961–97

Factories Act 1961
Construction (General Provisions) Regulations 1961
Construction (Lifting Operations) Regulations 1961
Construction (Health and Welfare) Regulations 1966
Construction (Working Places) 1966
Health and Safety at Work etc. Act 1974
Control of Asbestos at Work Regulations 1987
Management of Health and Safety at Work Regulations 1992
Workplace (Health, Safety and Welfare) Regulations 1992
Provision and Use of Work Equipment Regulations 1992
Personal Protective Equipment at Work Regulations 1992
Manual Handling Operations Regulations 1992
Health and Safety (Display Screen Equipment) Regulations 1992
Control of Substances Hazardous to Health Regulations 1994
Construction (Design and Management) Regulations 1994
Construction (Health, Safety and Welfare) Regulations 1996
Health and Safety (Consultation with Employees) Regulations 1996
Fire Precautions (Workplace) Regulations 1997

The Health and Safety at Work Etc. Act 1974

The Health and Safety at Work Etc. Act 1974 (the 1974 Act) is the kingpin of current legislation. It is an enabling Act, meaning that it enables health and safety regulations to be made from time to time without having to create fresh acts of Parliament. All the legislation listed in Table 6.2

subsequent to the 1974 Act (and much more) has sprung from this provision. It encourages a reasonably efficient response to be made to perceived new hazards (asbestos, for example, was once a new hazard) and changes in the law (such as European Directives). The 1974 Act set up the Health and Safety Commission (HSC) and Health and Safety Executive (HSE). It created new responsibilities for employers and employees, and provided for penalties.

The Construction (Design and Management) Regulations 1994

As their title suggests, the Construction (Design and Management) Regulations 1994 (CDM) are specific to the construction industry. Their purpose is to define a management structure for dealing with health and safety throughout every project, and to allocate responsibilities.

The onus for health and safety management initially falls upon the designer or designers, who must:

- ensure that their clients are aware of their own duties, before starting any design work for them
- ensure that the design does not expose to undue risk anyone associated with the construction, operation, maintenance, alteration and (eventually) demolition of the facility
- provide adequate information to those carrying out the construction about the design and materials used
- co-operate with the planning supervisor, who acts as a co-ordinator of health and safety requirements.

Uncertainty

As discussed in Chapter 4, the uncertainty of structural behaviour is managed statistically, using distribution curves. The equivalent tools for dealing with the more amorphous uncertainties of health and safety are:

- various modifying phrases in the legislation and the interpretation.
- advice provided by Approved Codes of Practice (ACOPs). ('Approved' means approved by HSE, signifying their status.)

Use of modifying phrases

Legislation is littered with phrases such as 'so far as is reasonably practicable', which define, as far as words can, the effort to be put into reducing health and safety risks, appropriate to the circumstances. These marker phrases have, as intended, acquired consistent meaning. In HSE's *A Guide to the Health and Safety at Work etc. Act 1974*, the term 'so far as is reasonably practicable' is defined as meaning:

> ... that the degree of risk in a particular activity or environment can be balanced against the time, trouble, cost and physical difficulty of taking measures to avoid the risk. If these are so disproportionate to the risk that it would be quite unreasonable for the people concerned to have to incur them to prevent it, they are not obliged to do so. The greater the risk, the more likely it is that it is reasonable to go to very substantial expense, trouble and invention to reduce it. But if, for example, the consequences and the extent of risk are small, insistence on great expense would not be considered reasonable. It is important to remember that the size or financial position of the employer are not taken into account.

Approved codes of practice

Each ACOP is linked to one of the regulations (although not every regulation has an ACOP) and explains and, where necessary, expands on the legal document. Although ACOPs are not law, they may be used in evidence. If harm has been caused, leading to prosecution, and if it is alleged that defendants did not properly discharge their responsibilities for health and safety, it will be sufficient for the defendants to demonstrate that the relevant ACOP was faithfully followed. If it was not followed, the onus will then be on the defendants to prove that they adequately discharged their duties in some other way. History tells us that such a defence is not often convincing.

Design and health and safety

Figure 6.1 (*overleaf*) shows a sequence of events that led to a fatal accident. The incident predated CDM; health and safety planning was at the time widely regarded as the contractors' sole responsibility, and there was obviously a failure on their part to take note of the proximity of overhead cables. Nevertheless, the accident might not have happened if:

- the excavation had been completed and backfilled before the scaffolding was erected, instead of afterwards; or
- the excavation had not been carried out so carelessly.

Now, post-CDM, questions might be asked about these trigger events. For example, if the excavation was done out of sequence because of late design, the repairer might share in the blame for the collapse. Liability would be greater if the repairer knew that the soil being excavated was weak and unstable (having previously done a ground investigation) but failed to warn the contractors. Finally, if the repairer inspected the work and failed to note that the excavation was not being shored up properly, then even more liability could be attracted.

The European Directive that led to the CDM Regulations was founded on the belief that most accidents can be traced back to flaws in design or organization. Although designers might challenge this assertion, most people would accept that accidents very often seem to spring from a background of poor health and safety culture, where everything from the working drawings to the site office looks untidy. It is commonplace to find, three-quarters of the way through a contract, operatives breaking out work that their colleagues finished halfway through, because the

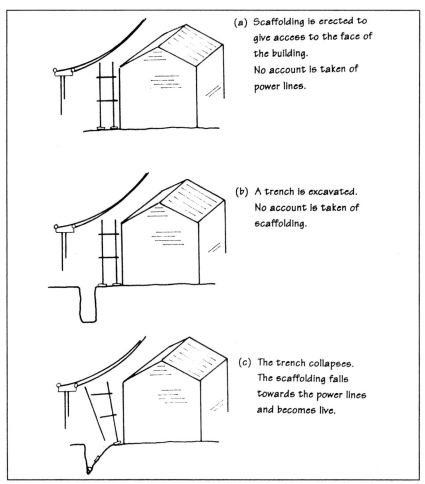

(a) Scaffolding is erected to give access to the face of the building.
No account is taken of power lines.

(b) A trench is excavated. No account is taken of scaffolding.

(c) The trench collapses. The scaffolding falls towards the power lines and becomes live.

Figure 6.1   Components of a fatal accident.

design has had to be revised or because the work was planned out of sequence. These errors arise from pressures that are peculiar to construction. Designers can improve the prospects of health and safety, in addition to discharging their formal duties, if they produce schemes that are:

- simple but comprehensive in detail
- buildable
- not subject to frequent change.

In fairness to designers, 'so far as is reasonably practicable' should be added, since many design flaws are born of impositions from without rather than errors from within.

Risk management

There is more than one way of managing risks to health and safety at the design stage. The Appendix A contains a simple, four-step method of risk assessment, appropriate for most simple projects:

- Identify every hazard.
- Rate its severity (1, 2 or 3).
- Rate its likelihood (1, 2 or 3).
- Calculate risk as severity times likelihood, leading to a value between 1 (trivial) and 9 (extremely high).

Before concluding this very brief review of health and safety design, it is worth recalling that the designer's duty to reduce risk to health and safety extends (under CDM) beyond the construction phase. Risks to users, cleaners, maintenance workers and subsequent construction workers must be considered. When, for example, designing substantial repairs to a flat roof, it might be timely to consider first installing permanent edge protection for the sake of future maintenance workers – especially if frequent visits are expected (for instance to a plant room). If the client can also be persuaded to replace fragile roof lights (often difficult to distinguish from the rest of the roof surface after a period of weathering) with safe alternatives, that would be a welcome bonus.

Table 6.3 (*overleaf*) lists hazards common to repair work. Some of these hazards are magnified when the work has to take place within a building. A few of the more unwelcome magnifiers are listed in Table 6.4 (*page 64*).

Table 6.3   Common hazards

Hazard	Tasks incurring hazard	Possible design alternatives
Asbestos *(causing asbestosis, lung cancer, mesothelioma)*	Disturbing or removing; pipe lagging, fibreboard, partitions, casing to services and steelwork, ceiling tiles, artex (pre-1985), underground ducts, sprayed fire and acoustic protection, jointing and caulking, rope fire stops. Digging up buried waste.	Carry out survey and follow standard procedure in The Control of Asbestos at Work Regulations.
Lead *(lead poisoning)*	Smelting, burning, welding, blast cleaning, demolition of lead and lead painted materials, disc cutting or using abrasive tools on lead, spray painting with lead based paints, working inside petrol tanks.	Avoid risk altogether if possible. Otherwise follow standard procedure in The Control of Lead at Work Regulations.
Fire *(risk of death)*	Hot working, smoking, burning	If possible avoid specifying flammable material (e.g. use fixings not adhesives) and avoid tasks involving hot working.
Confined spaces *(toxic fumes, suffocation)*	Any work in pits, tanks, flues, underground boiler houses, ducts, manholes, sewers, excavations and small rooms with poor ventilation.	If possible avoid or reduce time when operatives are at risk (e.g. use piled underpinning not hand-dug methods; line pipes rather than repair them; specify work that enables much of assembly to take place above ground).
Noise *(damage to hearing)*	Breaking out, drilling, driving.	Avoid if possible (e.g. use a self-levelling screed, if suitable, in preference to replacing existing screed); avoid lengthy operations.
Dust and fumes *(poisoning, respiratory disorders, eye irritations, silicosis)*	Breaking out concrete, cleaning buildings (especially siliceous masonry), cutting wood, burning, applying solvents, using internal combustion engines.	Refer to EH40 (available from HSE) for level of risk and consider less hazardous materials or methods (e.g. avoid chasing and cutting if surface fixing is feasible) or arrange for hazardous work to be done separately when building is otherwise empty.

Table continues

Table 6.3 continued

Hazard	Tasks incurring hazard	Possible design alternatives
Demolition *(falling, being crushed)*	Any work that potentially disturbs equilibrium or causes fall of materials or involves working at heights.	Identify features essential to stability and integrity. Identify uncertainties (e.g. unknown strength of floor after covering removed, size of bearings). Identify features needing particular care in removal (large trusses, continuous beams, prestressed concrete). Identify hazards not obvious to operatives (e.g. rotten floors, decayed masonry piers, dislodged structural members).
Contaminated land *(from fire and explosion)*	Any tasks involving skin contact or inhalation of contaminants.	Ensure that full information on contaminants is obtained and made available. Specify tasks that avoid or reduce contact with contaminants (e.g. piling rather than deep digging; if possible, piling that avoids bringing spoil to the surface; using a common trench for as many services as possible).
Excavation *(falling in, being buried)*	Hand-dug underpinning, working in deep trenches.	Ensure that all information on ground conditions is available. Identify existing buildings, services, pavings at potential risk during excavation. Prefer machine operations to hand digging at depth.

Table 6.4   Hazard magnifiers during structural repair

Hazard	Conditions causing magnification
Any activity	1) Sharing working space with other trades. 2) Sharing building with other users. 3) Working in confined and awkward spaces. 4) Unknown structural conditions e.g. floor strength. 5) Unknown, possibly hazardous, materials. 6) Possible difficulties for emergency access.
Falling from heights	Access difficulties encourage working from ladders instead of platforms.
Noise	Confined spaces and echoes in the working area, and transmission through the structure to parts well beyond the working area.
Dust and fumes	Poor ventilation, and spread through unsealed floors and walls, or ducts, to parts remote from the working area.
Manual handling	Difficult access for machinery increases the need for manual handling, which may also be in cramped conditions.
Excavations (collapse)	Proximity of existing foundations.
Fuel storage	May be necessary to store indoors or in less than ideal locations.
Fire	Storage of inflammable construction materials and accumulation of debris.

Chapter Seven

Diagnosis

Diagnosis should be carried out as methodically as design. There is no reason why it should not be as reliable, provided the person making it has a thorough knowledge of the behaviour of materials and gathers enough information about the damage to be in a position to distinguish true causes from potential causes. This chapter points out some common mistakes in diagnosis and recommends a method for generating the right answers.

Sherlock Holmes described the process of diagnosis thus:

> It is an old maxim of mine that when you have excluded the impossible, whatever remains, however improbable, must be the truth'. (Sir Arthur Conan Doyle)

His technique was effective – Holmes always got the right answer. Unfortunately, in the world of defect investigation, wrong answers still crop up occasionally, because many building professionals find the process of elimination difficult or frustrating. The most common mistake is the misapplication of simple *if…then* logic. For example:

> If this building were to settle, a tapering crack would form.
> A tapering crack has formed.
> Therefore this building has settled.

Of course, the appearance of a tapering crack admits settlement only as a possible cause, one of perhaps several causes that can form tapering cracks, so the last step of the argument should have been:

> Therefore settlement is a potential cause of damage.

This and all other potential causes should then be examined against as many indicators (or tests) as possible. For example:

> If this wall were to settle, its brick bed courses would be off level.
> The brick bed courses *are* off level.
> Therefore settlement remains a potential cause.

The cause versus indicator matrix

Figure 7.1 summarizes the mental process. Figure 7.2 catalogues the sort of factual information that yields useful indicators, although few investigations would need to trawl through the entire list.

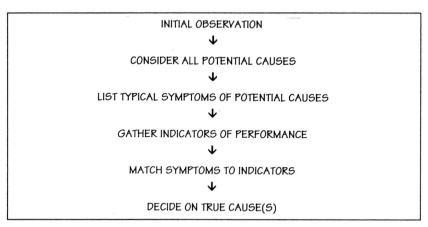

INITIAL OBSERVATION
↓
CONSIDER ALL POTENTIAL CAUSES
↓
LIST TYPICAL SYMPTOMS OF POTENTIAL CAUSES
↓
GATHER INDICATORS OF PERFORMANCE
↓
MATCH SYMPTOMS TO INDICATORS
↓
DECIDE ON TRUE CAUSE(S)

Figure 7.1   Steps to diagnosis.

Damage scale and pattern	Strains (compared with potential causes, likely behaviour).
Distortions	Brick bed course levels, other 'horizontal' features, vertical profiles, floor contours, any indicators of deflections, all differences between original and alterations.
Soil properties	Strength, modulus, stability, plasticity, water content, water table, redeposition, contamination, variations, mixes and layers and, above all, variations.
Chronology	Age of property, date of alterations, change of ownership, onset of damage, apparent progress. Environmental changes (dates of pumping, piling, development, vegetation – life and death).
Modelling the structure	Likely load paths through main and secondary elements. Evidence of changed load paths.
Environment	
Topography	
Records	Construction and alteration, soil conditions, previous land use, previous vegetation.
Position and condition of services	
Strength tests	
In situ load tests	
Chemical tests	
Establishing critical structural details	
Monitoring	Should not be allowed to delay diagnosis unnecessarily.

Figure 7.2   Factual information for yielding indicators.

POTENTIAL CAUSES	INDICATORS				
	Damage Pattern	Level of Damage	Distortion Survey	Soil Properties	Possible Agent
Settlement Under Load	✓	✓	✓	x	–
Thermal Strain	✓	x	–	–	–
Drying Shrinkage	x	x	–	–	–
Subsidence	✓	✓	✓	✓	x
Heave	✓	✓	✓	✓	✓

Figure 7.3   Potential causes and indicators.

Figure 7.3 shows a very simple matrix of causes versus indicators. A potential cause that is compatible with several indicators will be a promising candidate for 'true cause', but no amount of testing will provide absolute proof. On the other hand, a single non-compatibility will disprove the potential cause. For example, in the hypothetical case on which Figure 7.3 is based, settlement under load gets three positive indicators, but is disproved because it does not satisfy the 'soil properties' test. The soil may be far too stiff and strong to have allowed settlement damage to form. Soil properties are irrelevant to thermal damage and drying shrinkage, so this indicator provides no evidence for or against those potential causes. It does apply to both subsidence and heave and, on this occasion, admits both. Finally, only one cause – heave – emerges without any crosses, and that is therefore announced as the sole cause of damage.

It is not usually necessary to produce a formal 'cause versus indicator matrix' with a report on causation, as long as the mental process has been followed. A printed matrix would, in fact, be cumbersome. An experienced investigator would, either mentally or on paper, use more than the five generic indicators shown in Figure 7.3, even for a simple case. *Structural Appraisal of Traditional Buildings* (see *Further reading*) gives examples of applying and interpreting indicators.

The matrix will produce the right answer for simple cases with little fuss, but sometimes, with the more intractable cases, the tick or cross check may be too crude to be infallible. Shades of opinion may need to be substituted for the ticks. These may be expressed verbally ('unlikely', 'likely', 'very likely') or numerically, using a likelihood ratio, so that the cumulative strength of several indicators can be judged. Figure 7.4 (*overleaf*) shows a house whose sagging roof was strengthened and stiffened by the insertion of trusses, creating a redistribution of roof loading. Immediately after the work was completed, some cracking in the front and flank walls appeared. One potential cause was settlement arising from load redistribution; another was quite unconnected with the work on the roof.

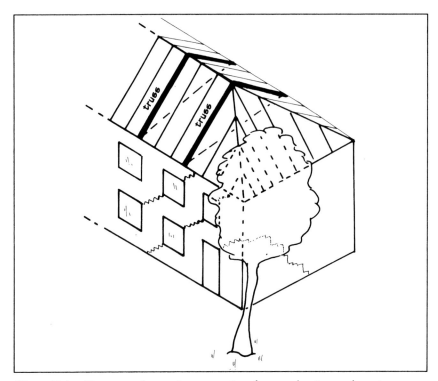

Figure 7.4   Damage: alternative causes (roof strengthening and tree).

Figure 7.5 shows how the likelihood of the two rival causes might have been judged (without reference to the lengthy relevant data and analysis).

In some cases, there is more than one cause of damage. They may or may not have equal weight. If it is important to obtain as precise a cal-culation of relative influence as possible, which could be the case where potential repair costs are more sensitive to one cause than the other, then load testing and monitoring might be added to the suite of indicators in the likelihood test.

Even with the cleverest system, it is difficult to spot an incipient cause, especially if its symptoms have yet to emerge from background noise. Figure 7.6 shows a building that subsided when an influx of water eroded the made ground beneath one of its corners. This was correctly diagnosed. The problem was solved by underpinning to the top of the clay beneath the made ground, but that stratum was already beginning to heave because it was recovering from a previous desiccation. The heave symp-toms were, at the time of the investigation, masked by the more vivid symptoms of subsidence. Heave was simply not recognized. The repairs, unfortunately, were inappropriate because they were soon defeated by the continuing heave movement.

It is possible to apply the cause versus indicator matrix of likelihood testing flawlessly, yet get the wrong answer because of shortcomings in technical knowledge. For example, slope stability is a difficult subject to master. If those carrying out the diagnosis have limited knowledge, they

INDICATOR	POTENTIAL CAUSES	
	Roof Stiffening (Causing Settlement)	Horse Chestnut Tree (Causing Subsidence)
Case history	Very unlikely	Likely
Location of cracks	Unlikely	Likely
Pattern of cracks	Unlikely	Likely
Level surveys	No indication	Likely
Maximum & differential foundation movement	Very unlikely	Likely
Damage: onset	Likely	Very likely
Damage: progress	Very unlikely	Very likely
Soil plasticity	No indication	Permits this cause
Presence of roots beneath foundations	No Indication	Permits this cause

Figure 7.5   Likelihood test.

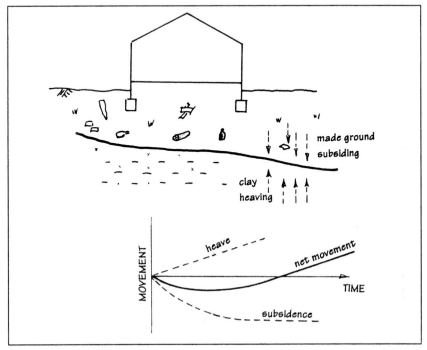

Figure 7.6   Symptoms difficult to distinguish.

may not recognize the indicators that scream 'slope instability is not a possible cause'; slope instability will then survive the scrutiny and the remedy for it may cause mischief. It is also possible, because of gaps in our knowledge, to omit potential causes altogether. For example, most brickwork is clay, and clay brickwork does not undergo drying shrinkage. The investigator may know that clay bricks do not shrink, but may be unaware that:

- not all bricks are clay
- some are calcium silicate
- calcium silicate bricks do shrink quite noticeably, unless movement joints are carefully detailed.

If that is the case, a potentially important indicator may be omitted or misapplied. Diagnosis may be wrong. Repairs may be inappropriate.

Enquiry can be no more rigorous than the knowledge that guides it. The pillars needed to support the logic of diagnosis are:

- knowledge of building materials
- knowledge of structural behaviour and soil behaviour
- a repertoire of indicators or tests
- specialist advice when necessary
- a sense of proportion.

A final word on indicators: experienced investigators quickly decide how to deploy their battery of possible indicators to greatest effect without wasting time and cost. It is not a task that should be done by checklist, because every case is different and needs its specific set of indicators. Initial inspection of the building by an experienced investigator is essential. Data produced by indicators loses impact when it is generated by checklist and interpreted by someone who has never seen the building.

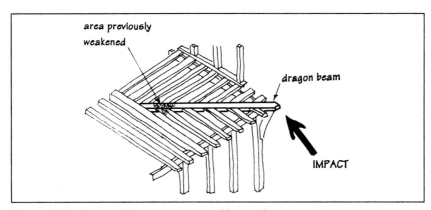

Figure 7.7   Impact damage aggravated by weakness.

Trail of suspicion

On an urgent visit to appraise, say, vehicle impact damage, an investigator will be excused the full Sherlock Holmes diagnosis. It would, of course, be useful to have sufficient knowledge of structural behaviour to assure the building's immediate safety. Furthermore, any trail of suspicion should always be followed. Figure 7.7 shows the case of a medieval jettied building that was struck by a delivery lorry. The impact occurred at an external corner of the upper storey, which was supported on a substantial dragon beam. Shop assistants working in the ground floor jewellers described it as no more than a jolt, and business continued as usual, but damage proved to be disproportionately serious. Moreover, permanent distortions indicated there had been less rebound than usually happens when healthy timber buildings are struck. The trail of suspicion led to the dragon beam, which had been previously weakened by water leakage inside the building. The impact had been energetic enough to fracture the weakened section, leaving the first floor only marginally stable. If cosmetic repairs only had been carried out, as at first sight seemed appropriate, the building might later have been seriously damaged by normal use.

Handicaps

Coverings, decorations, budgets, and even cunning disguise all conspire to curtail investigation. Incomplete information is a cross investigators will always have to bear (Chapter 27).

PART TWO

Defects

Chapter Eight

Causes of structural damage

This chapter provides a checklist of the most common causes of damage. Other chapters in Part Two expand on the more serious and more progressive of these.

Common causes of damage

Foundations

Settlement under load.
Subsidence: erosion by water.
Subsidence: inundation (collapse compression).
Subsidence: shrinkage of clay or peat.
Subsidence: pumping and ground water lowering.
Subsidence: vibration of granular subsoil.
Subsidence: swallow holes.
Subsidence: shallow mining or abandoned shafts.
Subsidence: deep mining.
Subsidence: consolidation of made ground under self-weight.
Heave: swelling of clay.
Retaining wall failure.
Reducing external ground level.
Slope instability.
Soil creep.
Loading of adjacent ground.
Chemical attack.

Note: The pattern of damage is greatly influenced by the type, strength and stiffness of the foundations. Defects in the foundations, whether or not inherent, may exaggerate the damage and will influence its pattern.

Ground floor slabs

Consolidation of sub-base	Loose hardcore, organic material.
Heave of sub-base	Sulfate expansion within the hardcore.
Frost heave of silty, chalky or sandy soils	Rarely a problem in occupied buildings.
Defects in screed	Weak or variable material; inappropriate thickness or preparation of substrata.
Flood	Rarely a structural problem, except in basements, where the floor may lift if hydrostatic pressure develops.

Note: Ground movements affecting foundations can also affect the ground floor.

Timber floors

Shrinkage	Initial drying; central heating; reoccupation.
Overload	Especially heavy partitions built off joists.
Rot	Especially where ventilation is inadequate.
Infestation	Encouraged by damp.
Alteration	Notching for services; removal of partitions; change of use introducing heavier loads.

Internal partitions

Shrinkage	Initial drying; central heating; reoccupation.
Loss of support	Settlement, shrinkage or deflection of supporting member.
Alteration	Cutting holes can damage the ability of a partition to act as an arch or truss, or as restraint to load bearing walls.
Overload	Especially if the partition relies on supporting beam or joists.
Concentrated loads	Causing cracking or spalling at beam or lintel bearings.
Loss of support	Causing gross distortion where load bearing elements have been removed

External walls: masonry

Foundation movement	See under *Foundations*.
Thermal movement	Especially south facing walls longer than 12m.
Irreversible moisture movement	Shrinkage of calcium silicate or concrete units; initial expansion of clay brickwork.
Frost	Especially soft bricks in hard mortar.
Weathering of mortar	Often preceded by frost attack; sometimes followed, in cavity walls, by loss of bond of wall ties.
Deterioration of units	Frost, solution, pollution, salt crystallisation, decay of reinforcing bands (metal or timber).
Rubble consolidation	The loose core may consolidate and cause facings to bulge.
Rusting	Principally of wall ties, but also bands of steel or iron reinforcement, iron clamps and inserts in older buildings.
Masonry bee attack	Removing mortar.
Sulfate attack	Causing deterioration and expansion.
Wind	Rarely a problem below eaves level, except where there is a combination of slenderness and deterioration.
Concentrated loads	Causing cracking or spalling at beam or lintel bearings.
Loss of support	Causing gross distortion where load bearing elements have been removed.
Delamination	Separation of poorly tied leaves causing bowing.

Note: In old buildings, original connections between external walls and cross walls, floors and roofs were often weak, but were adequate until reduced by deterioration or minor movement, leading to a hidden increase in slenderness, and increased vulnerability to damage caused by movement or loading.

External walls: timber

Foundation movement	See under *Foundations*.
Shrinkage	The most common cause of medieval distortion (usually complete).
Rot	Large sections may remain adequate, but rot of joints and plates can lead to disproportionate damage.
Infestation	Rarely absent from medieval buildings; rarely a problem unless active.
Alterations	Especially removal of ties and struts whose function was not understood. Previous alteration, such as installation of windows under beams, can create sensitive spots, but this is not often a severe problem.
Wind	Rarely a problem, except where a weakness has been introduced.

Note: In medieval timber frame buildings, connections between posts and other elements (plates, beams, joists, bracing and infill or cladding) may be essential to stability. Local rot and apparently minor alteration can make the external wall suddenly more vulnerable to damage caused by movement or loading.

External walls: earth

Damp	May cause weakness – severe if flooded.
Frost	May combine with damp to cause weakness or reduce sections.
Vermin	Rat runs may reduce sections.
Overload	Alterations may overestimate the strength of the wall.
Wind	Storm damage to weakened walls.

Note: Earth walls are weaker than other materials, and their connections with cross walls, floors and roofs are often feeble. Damp, frost and vermin can cause a sudden and catastrophic loss of strength.

Gables, chimneys and parapets

Wind	Suction of pots and top brick courses, rotation of parapets.
Frost	Exposed elements, which are then more likely to suffer wind damage.
Thermal	Excess heat in unlined chimneys; diurnal variations causing shunting of parapets.
Sulfate	Especially on chimneys, creating a leeward lean.

Roofs

Wind	Especially uplift of flat or low pitched roofs.
Snow	Especially valleys where it can collect.
Overload	Replacement of roof covering by something heavier; storage in roof space; new tanks.
Spread	Horizontal component of rafter load not resolved at plate level.
Thermal movement	Flat roofs, especially concrete.
Alteration	Removal of struts and collars to provide living space; altering trussed rafters; misplaced additions; re-covering using heavier materials.

General

Deterioration	May or may not be accelerated by aggressive environment, incompatible materials, poor maintenance and repair and abuse.

Fire
Impact
Blast
Dust explosion
Lightening/fireball strike
Vibration
Alterations

Note: An unexpected increase in load, even if it is not abnormal, may expose the weakness of understrength members or connections.

Chapter Nine

Below-ground defects

Most foundation movement is progressive. But not every case is either serious or long term. This chapter explores the range in severity and time of the various causes listed in Chapter 8, under Foundations.

Settlement under load

On coarse soils (defined in Chapter 3 under *Particle size*), settlement is nearly complete when the building is complete. On fine soils, initial settlement is followed by a period of consolidation as water drains out under pressure (Figure 3.4, *page 26*). This period can be weeks or months, sometimes years, depending mainly on the permeability of the soil. There is a final 'creep' phase which, in most natural soils, is too insignificant to cause even cosmetic damage. In this context, creep can be described as movement at an ever-decreasing rate. A rough approximation is obtained by plotting a straight line for movement against a $\log_{10}$ timescale. This means that the first year of creep sees as much movement as the next ten,

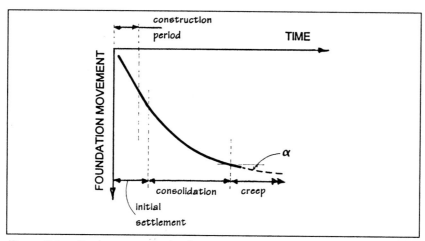

Figure 9.1   Settlement under load.

and the next hundred. As with all structural distortions, movement is not steady, but consists of occasional stop–start actions; creep would have to be monitored over a long period (often an impracticably long time as far as a repair contract may be concerned) to obtain useful information.

Figure 9.1 shows a standard settlement curve for soils with measurable movement in all three phases. (Gravel would virtually stop moving at the end of initial settlement.)

Made ground will exhibit all phases of settlement, including creep, unless it is inert and has been well compacted. Table 9.1 gives a rough indication of the time taken by a variety of materials to consolidate after deposition. In some cases, this can be used as a preliminary rough guide to judge whether repairs to a building settling on made ground can be simple and superficial, or whether they need to include underpinning. If the difference in cost between substantial and superficial repairs is large, further investigation should be carried out, including testing and, if time permits, monitoring.

Table 9.1   Consolidation and creep of made ground

	Typical time for 90% consolidation (*years*)	Consolidation as a % of original thickness	Angle of creep line to the horizontal (α)
Coarse soil well compacted as it was placed	1 – 2	1% – 2% Less if well graded	0.2%
Coarse soil simply tipped	3 – 5	3% – 5%	0.5%
Coarse soil tipped in water	3 – 5	4% – 8%	0.5%
Rock fill with some compaction	3 – 5	3% – 5%	0.5%
Well compacted rock fill	1 – 2	1%	0.2%
Open cast mining fill, uncompacted	5 – 10	4% – 8%	0.5% – 1%
Clay with some compaction	5 – 7	8%	1% – 2%
Tipped clay	7 – 10	12%	2% – 3%
Lagoon placed clay	10 – 15	15%	3% – 5%
Tipped chalk	8 – 10	12% – 15%	2% – 4%
Chalk with some compaction	6 – 8	10% – 12%	1% – 3%
Domestic refuse	5 – 50	30%	3% – 10%

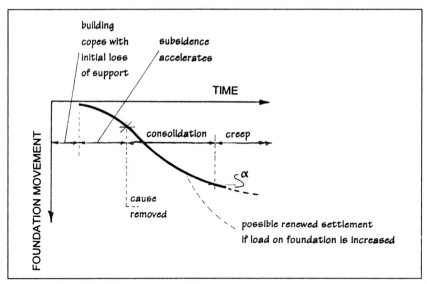

Figure 9.2   Subsidence: escape of water.

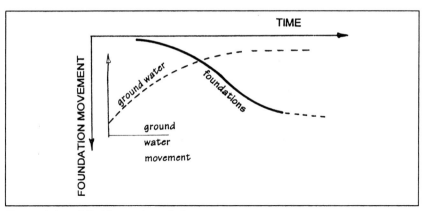

Figure 9.3   Subsidence: inundation.

Effect of water on coarse soil

There are two common mechanisms by which coarse soil and made ground can be destabilized:

- *Erosion by escape of water* tends to be local to the cause, and is usually remedied by stemming the source.

- *Inundation by influx of water or rise in water table*, which affects only made ground, tends to affect the whole building, and the cause cannot always be controlled.

Erosion by water

Silt and sand may be eroded if water travels through the soil with sufficient energy to dislodge individual particles and carry them at least a short distance away. This increases the ratio of void to solid within the soil, and reduces the soil's angle of friction. At a critical point, internal friction between soil particles becomes insufficient to support the soil's own weight and any applied load. The particles slide across each other. In so doing, they pack down to a more stable arrangement. Soil volume is thus reduced; and, of course, volume reduction below foundation level causes subsidence. Figure 9.2 shows a typical record of foundation movement.

Inundation

Made ground that is poorly compacted or exceptionally dry can sometimes consolidate rapidly when large volumes of water enter it. The process is termed 'collapse compression'. Mining waste, chalk and redeposited clay are particularly vulnerable. The water may come from below, in the form of a general rise in the ground water level, or from above as rainwater percolating downwards from soakaways or leaking drains, or from a burst water main.

Figure 9.3 shows a typical record of movement. Problems have arisen from as little as a two-metre rise in ground water level. Designed foundations, even piling, have not always been immune. Because the cause is not triggered until water is supplied, buildings of all ages have been affected, even historic buildings. Subsidence movements of up to 7% of original thickness have been recorded.

As indicated in Figure 9.3, movement may continue after the water table has stabilized. The repairer would have to take into account this uncertain continuation and the risk of a second incident of inundation. Repeat inundation rarely leads to more than a fraction of the damage caused by the first occurrence. However, if the rise in ground water level is greater the second time around (Figure 9.4, *overleaf*), then it would be the first event affecting newly inundated material, and the new damage would in some circumstances match the original in severity.

Vegetation and clay

For simplicity, this section will discuss vegetation in terms of a single tree. However, in actual cases, there are often two or more specimens, sometimes difficult to distinguish from each other; and large shrubs and climbers can, of course, also cause serious problems.

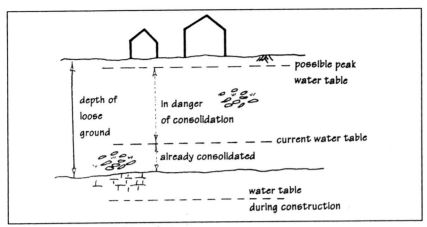

Figure 9.4   Potential inundation.

No two cases are the same, but Figure 9.5 attempts to illustrate the generic problem. It plots the progress of subsidence caused by a single vigorous tree in a highly shrinkable soil of low permeability, followed by recovery after tree removal. In this imagined case, early shrinkage (stage A) is not enough to cause measurable movement of the foundations but, as the water demand of the tree increases during summer months, shrinkage begins to make its mark on the building. Winter months see the clay recover its moisture, and swell, and the foundations move back. This is stage B. (In practice, recovery would not be complete in abnormally dry winters, but basically movement remains seasonal during this stage.) As summer water demand continues to increase, a persistent moisture deficit begins to form within the clay (stage C). A certain amount of winter recovery may still occur from time to time, but from then on the soil never returns to its base, or 'pre-tree', condition. Subsidence continues, worse in some years than others, although it tends to level out after a few seasons, once the point has been reached where the tree's demand for water reaches a rough equilibrium with the soil's ability to supply it.

Figure 9.5 assumes that the tree is eventually removed, precipitating recovery (stage D). The rate of recovery is driven by the negative pore pressure (suction) that has developed within the clay as the tree has been withdrawing water. The higher this negative pore pressure happens to be, the more readily it will absorb moisture percolating downwards from rainfall or migrating (draining) inwards from the immediately surrounding wetter soil. This mechanism has been described in Chapter 3, and illustrated in Figure 3.5 (*page 28*). As the water content is replenished, suction reduces and the process tends to slow down.

If the water supply were unlimited, recovery would follow a steady curve whose inclination would be determined by soil permeability (Figure 9.6).

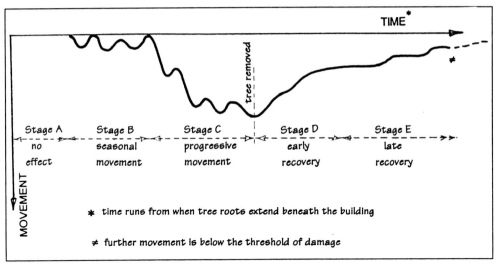

Figure 9.5   Clay–tree subsidence.

A steep curve denotes high permeability, perhaps that of a sandy boulder clay, and a flat curve denotes very low permeability such as we might expect in one of the overconsolidated clays. Towards the end of the process, the difference in pore water pressure between tree-affected and non-tree-affected parts of the soil is too small and uneven to sustain a relentless drive towards further recovery, and progress is then at the mercy of the weather. In the UK, there are relatively short periods – a few days every so often – during which there is enough rainwater, after run-off (surface drainage), evaporation and transpiration, to fuel recovery. Late recovery (stage E in Figure 9.5) therefore takes place in fits and starts. During a dry period, little or no movement will take place, and this may give the false impression that stability has been achieved.

The following additional comments (referring again to Figure 9.5) may be a helpful, if brief, commentary on the bare outline.

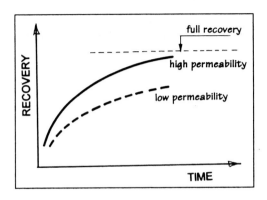

Figure 9.6
Recovery and permeability.

Stage A

Most tree root systems are shallow, with the majority of roots in the top 600mm and few deeper than 1000mm. Suction will cause some desiccation below the root line. Foundations of reasonable strength and stiffness will adjust to the modest redistribution of bearing pressure entailed by minor suction, with little or no impairment. We can say, therefore, that most buildings with foundations up to minimum (treeless) Building Regulations standards will live in harmony with trees, even trees that are growing closer to the building than their height. (Of course, 'most' does not mean 'enough'. NHBC guidelines for buildings near existing trees should always be the minimum standard for new construction.)

Stage B

Many cases remain indefinitely in the seasonal stage. That is especially so with older buildings near modest vegetation. These older buildings often have shallow foundations that are well within the suction zone of typical tree root systems.

   Seasonal subsidence continues for as long as the soil is able to meet the demands of the vegetation. If and when it cannot, for whatever reason, the roots of the vegetation will need to expand, horizontally or vertically or both, if the vegetation is to remain healthy. The trigger event is often the increasing water demand of a maturing tree; the crucial factor in determining whether the soil can meet this demand without expansion of the root system is soil permeability. Impermeable clays, such as London clay and other overconsolidated clays (typically very highly plastic) take the longest to recover from summer deficits, because their low permeability delays winter recovery. When replenishment falls behind, a persistent deficit is established and progressive subsidence is initiated. More permeable clays, such as many of the sandy or chalky boulder clays, alluvial clays and soil mixes that have clay as a secondary constituent, are more likely to recover every year, and therefore subsidence is likely to remain seasonal. (Drought conditions can lead to increased subsidence, especially in areas that normally receive average to high rainfall, but this is usually a short-term problem, and not progressive.) Once again, unfortunately, there is no means of predicting individual cases with certainty.

Stage C

This is the stage in which prediction of future movement would be most useful, but is most difficult. In addition to the general imprecision of tree and soil behaviour, in stage C we are often dealing with extreme cases

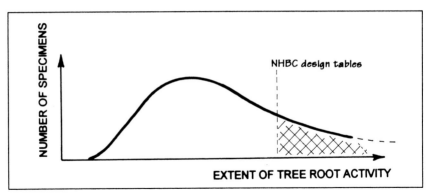

Figure 9.7   Distribution curve: tree root activity.

of root spread, analogous to characteristic values being exceeded (Chapter 4 and Figure 4.5, *page 48*). If distribution curves were produced for the root activity of each species of tree (and no doubt every species would be different), they would be much flatter than the typical material strength curve. Figure 9.7 shows a flat distribution curve, on which 'extent of tree root activity' can be read as either radius or depth. The sketch is intended as only a crude illustration that, when subsidence is progressive (stage C), the vegetation is likely to have grown into the tail of the distribution curve, and desiccation may be excessive. Notwithstanding the uncertainty implied by Figure 9.7, there is a tentative hierarchy of trees' potential for causing subsidence. Table 9.2 (*overleaf*) shows where a few of the more common UK trees would fit within this hierarchy, relative to each other, assuming a common distance between tree and building (say, 10m or so – but it would depend on the species of tree). Actual risk, in terms of likelihood of subsidence, would be difficult to forecast, even with comprehensive data on all the conditions that influence it (some are mentioned in the table). The implications of all this uncertainty are discussed in Chapter 25.

Stage D

Figure 9.5 shows recovery beginning with the felling of the tree. If the tree loses vigour and dies naturally, there may be a long transition between stages C and D.

The repairer would like to know how much recovery to expect and how long it will take. This is discussed in Chapter 3. Unfortunately, both predictions are elusive. The experienced professional should have a 'feel' for whether the particular recovery under consideration is likely to be weak, moderate or strong; a period of monitoring may improve the accuracy of the feeling, but there are no precise answers. (If total previous subsidence,

Table 9.2   Trees' potential for causing subsidence

This table lists the potential for subsidence (from high to low) of 24 common species. See also NHBC regulations and BRE Digest 298 (current edition).

Poplar	Conditions affecting actual risk
Crack willow, white willow	
English elm	*Condition of tree:*
	Root damage
Red oak, turkey oak	Branch damage
English oak	Pruning, thinning, etc.
	Environment:
Weeping willow	Urban/rural
Wych elm	Inland/coastal
Holm oak	Shade and shelter
	Rainfall
Leyland cypress	Pollution/contamination
Other cypress	*Ground surface:*
Plane	Topography
	Covering if any
Sycamore	Drainage if any
Lime	*Below ground:*
Ash	Obstructions
	Trenches
Horse chestnut	*Soil:*
Wellingtonia	
	Shrinkability
Walnut	Permeability
False acacia	Acidity/alkalinity
	Various mixes and layers
Cherry	
Cedar	Advice on specific risks should be
	taken from suitably experienced
Beech	professionals: arboriculturist can
Birch	advise on tree behaviour and civil/
	structural engineer can advise on
Holly	building performance.

Note: Every tree has a preferred depth of rooting. Environmental conditions (see above) may challenge the preferred pattern. Trees will attempt to adapt to their environment, which may itself change, e.g. through drought. Some trees can adapt more successfully than others. Hence the order given is not fixed for all conditions.

from zero time in Figure 9.5, can be estimated with reasonable approximation, this would be as good an indication of recovery as any.)

Stage E

Even when monitoring (or simple observation) confirms a period of
apparent stability, we cannot be certain that the recovery is complete.
There have been many cases of buildings suddenly manifesting recovery (or heave) damage after some years of quiescence, although at that
late stage of the sequence the level of damage is usually only cosmetic.
Before measurable movement stops altogether, it fades away like creep
settlement (Figure 9.1, *page 80*), eventually falling below the rate at which
it causes even cosmetic damage.

Building performance during recovery

The level of damage caused per unit volume of clay expansion (in other
words, how easily the building is harmed) is influenced by:

- differential pressure of the expanding clay (differential because the suction that engendered it was inevitably differential)
- the ratio of soil stiffness to structure stiffness (as discussed at the begin-
  -ning of Chapter 4)
- the strength of the building's load-bearing walls.

These factors, or similar ones, influence the course of all below-ground
defects, but two features are peculiar to clay recovery:

- Expanding clay tends to produce an abrupt variation in soil pressure.
  Most other ground movements have gentle variations in comparison.
- Expanding clay is stiff; the soil–structure stiffness ratio is higher than
  it would be for the same building on most non-plastic soils. This
  reduces the capacity for beneficial soil pressure redistribution. In other
  words, the building has limited opportunity to mitigate the damage.

If the third factor is also discouraging (if load-bearing walls are not strong),
we can expect the building to be sensitive to damage. For example, a one
hundred-year-old single-storey building of masonry or earth material
would certainly be vulnerable to small movement; case history suggests
that such buildings can be damaged by maximum movements of as little
as 5mm in a season. Stiffer, stronger buildings obviously fare better, but
it is not possible to give accurate limits for various types. As for many other
predictions, experience is the best guide. Two obvious points may be made:

- A building with well-defined fractures will offer relatively little resistance to recovery. That is beneficial during stage D recovery, because it encourages existing cracks to close in preference to new cracks opening. However, repairs cannot be delayed for ever and, at some point in stage E, it will have to be decided whether to let crack closing continue or carry out repairs that will improve resistance to the residual movement.
- Whereas it can be expensive to arrest small movement, it is often relatively easy, by judicious strapping, to improve resistance to the damage it causes (Chapter 17; Figure 17.4, *page 160*).

Long-term performance

Recovery is not quite the mirror image of subsidence. Pressures, distortions and damage do not unfold and fold in exactly the same way, mainly because the factors influencing performance (differential soil pressure; soil–structure stiffness ratio; building strength) change their values slightly as the circumstances change. Detailed discussion of this subject would yield diminishing returns because, once again, precise data are elusive. Suffice it to say that the result, at the end of a full cycle of subsidence and recovery, is usually a small net downward movement.

Recovery, then, usually falls a little short of subsidence. Buildings that undergo several cycles of seasonal movement may experience long-term subsidence by increments. Distortions may accrue and damage may magnify over a long period, rather like the ratchet effect sometimes caused by thermal movement. (Chapter 4 includes a brief discussion of redistribution of loading and bearing pressure, not specific to subsidence and recovery cases. Chapter 12 discusses thermal 'ratchet' movement.)

Heave

If the tree is removed before the house is constructed, the risk of subsidence will be eliminated but, unless the foundations are designed to cope with swelling soil, a risk of heave will be introduced. The building has stages D and E (Figure 9.5, *page 85*) to withstand. Obviously, the more time that has elapsed between tree felling and building construction, the less heave there will be. But, as always, prediction is difficult, the more so because there are no data from a subsidence phase. Heave can be more damaging than subsidence in the following circumstances:

- if the building is located on the site of the removed tree, where degree and depth of desiccation are likely to be greatest, and differential pressure during heave will be at its worst

- if the tree has caused maximum (stage C) desiccation – there is no opportunity during heave, as there usually is with subsidence, to mitigate damage by early action
- if lateral heave also develops (see the following section). In combination, vertical and lateral heave can be devastating.

Heave affects mainly young buildings. Before 1980, many builders were unaware of the problem. Moreover, products for cushioning or avoiding the movement (Figure 3.13m,n, *page 40*) had not been developed, so heave damage was common in areas of very highly plastic clay. It is now an infrequent cause of damage.

Lateral heave

Lateral heave damages buildings only if both of the following conditions are satisfied:

- Foundations extend their internal vertical face into the heave zone.
- The heave pressure on the internal face of the foundation is able to overcome restraint ('passive pressure') against the external face.

Figure 9.8 shows an example. It is probable that lateral pressure from clay swelling has developed within the perimeter of many substructures, but has not often matured into movement and damage, because it has been resisted by the soil lying against the outside face. Resistance can be thwarted by later events such as:

- roots from live vegetation causing the soil to shrink (as shown on Figure 9.8), permitting the foundations to slide into the gap that forms

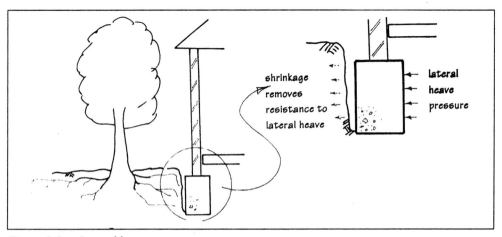

Figure 9.8   Lateral heave.

- seasonal water content changes causing shrinkage within 1.5 metres of ground level, again leaving a gap beside the foundations
- soil excavated locally.

These incidents do not necessarily coincide with the progress of any vertical heave. Lateral heave can happen in the absence of vertical heave, or after vertical heave has substantially run its course; or the two can be active at the same time, although not necessarily following the same timetable.

Anti-heave precautions

When resilient materials are used to protect modern foundations against heave (Figure 3.13m,n, *page 40*) they yield under the pressure of the expanding soil; by yielding, they diminish this pressure, but they do not eliminate it. A certain amount of pressure is always transferred through the resilient material to the foundations. If the design has been successful, the value of this transferred pressure should be too small to overcome dead weight and inherent strength. But thickness is critical. If the resilient material is no thicker than the potential free (unopposed) change in soil volume, then it would be completely ineffective. If it is only a little thicker, it would be of marginal use – it would effect a small reduction in pressure; and so on. Obviously, repairers must consider this when designing underpinning (Chapter 17). Less obviously, investigators should bear it in mind when diagnosing damage. It should not be taken for granted that the presence of anti-heave precautions rules out the possibility of either vertical or lateral heave altogether. (If there is doubt, a specialist should be asked to give advice based on all available indicators, including the type, thickness and actual performance of the anti-heave material.)

Clay heave without vegetation

A minority of heave cases cannot be linked to prior removal of trees or shrubs. A persistent source of water, from land drains or a forgotten ditch, or from a leaking water main in the building itself, can raise the clay water content locally and cause differential heave pressure. The water source has to be persistent, because the (usually) low permeability of shrinkable clay ensures that water from a point source can only be absorbed slowly.

A smaller number of heave cases, involving no vegetation and no persistent water source, are perplexing to the investigator. Usually, the only likely explanation is a long-term rise in clay water content, possibly associated with a previous land use. In one case, aerial photographs confirmed that the site had been farmland for thirty years before construction; it was remote from boundary hedges or any other form of vegetation, apart from crops – and yet monitoring demonstrated a long-term heave. Movement was continuing, albeit intermittently, when monitoring stopped after twelve years.

In certain urban areas, a rise in ground water level is feared as a result of declining industry taking less water from nearby boreholes. If this were to happen, buildings with steps in level (such as basements under part of the plan area) would be most at risk, if the subsoil is shrinkable clay.

Finally, clay heave can be generated by the removal of overburden. This affects mainly overconsolidated clays, rather than younger boulder clays and alluvial clays. Every metre of soil removed causes a reduction in pressure of about 20 kN/m^2; and there is a corresponding volume increase associated, as always, with drainage (Chapter 3). This time the volume change is predictable. It rarely causes serious problems, partly because the pressure is fairly uniform and therefore differential movement is small. There would be at least some pressure variation on sloping sites, but slopes are modest on overconsolidated clay. Nevertheless, ground floors are liable to lift if they are not protected from clay heave by a void or collapsible material beneath them. For that reason, basement floors should always be suspended if shrinkable soil is removed to form them.

Non-clay heave

Peat is very highly shrinkable and is certainly vulnerable to subsidence caused by tree root activity. Recovery is usually weaker than clay recovery (Chapter 3).

Although in the UK frost heave is not a common problem in inhabited buildings, prolonged sub-zero temperatures can cause silt and chalk to swell sufficiently to damage lightly loaded shallow foundations.

Pumping and ground water lowering

Ground water lowering is sometimes used to permit work on civil engineering and building sites to be carried out in dry conditions. In a minority of cases, this can cause subsidence to existing buildings within the radius of influence. Where this is expected, the project should include careful surveys of all buildings at risk, before and after ground water lowering, so that existing ('before pumping') damage can be agreed. If sensitive or historic buildings are at risk, the local planning authority should be consulted. If the risk cannot be eliminated, the buildings should be monitored. Piezometers may be used to monitor the movement of ground water at the site, to provide a check on predictions and early feedback for controlling the pumping. Unfortunately, these elementary preparations are occasionally neglected.

Figure 9.9 shows the effect on ground water of pumping from a single well. (In practice, a line or group of wells would probably be used.) A drawdown is created, whose shape depends on the permeability of the soil. In soil of low permeability, the curve is steep and the radius of influence is small. For that reason, pumping would not usually be considered in pure clay; but it is occasionally considered when clay is one component within mixtures or layers of soil. Figure 9.10 shows an example of ground water lowering in layered soil. If, in the layered case, pumping continues for a long period, the clay will start to drain because, although its permeability is very low, the alternating strata of coarse soil offer huge areas through which the water can migrate. As water drains out, the clay will shrink in the same way as it does when affected by loading or suction (Figures 3.4 and 3.5, pages 26 and 28).

Soil that was previously buoyant, because it was below ground water level, gains in weight by more than one-third when it is dried out. In theory, this can create fresh consolidation. In practice, most natural sands and gravels have sufficient reserves (are sufficiently dense) to be almost immune to subsidence by pumping, provided the method includes filtering to prevent the removal (erosion) of fine particles within the soil being drained.

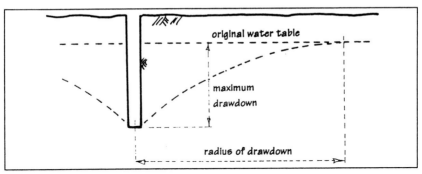

Figure 9.9   Ground water lowering.

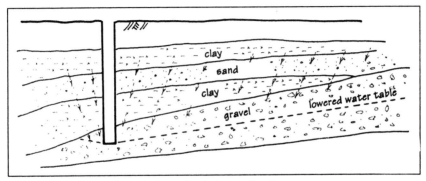

Figure 9.10   Ground water lowering in layered soil.

Pumping is not, therefore, inevitably a menace to nearby buildings, because many types of soil are insensitive to it, for one reason or another. The soils that are most vulnerable are: coarse silt and fine sand, especially if they are loose; peat, which readily loses a proportion of its enormous water content through pumping; made ground, especially if it is loose; soft clay, if its permeability allows a significant radius of influence to develop during the period of pumping; and layered soil, if a large radius of influence can develop quickly.

Vibration

Airborne vibrations rarely cause structural damage, although they may rattle windows and occasionally strip roof tiles when the source is low-flying aircraft.

Ground-borne vibrations can arise from construction sites, from traffic or from machinery inside the building. Unless it contains a generous proportion of granular material, clay is not sensitive to vibration. It may be affected if it is a constituent of made ground, or is a thin stratum among more vulnerable layers. Dense sand and gravels are affected only by very severe vibrations. Silt and sand, especially if loose, made ground, especially if loose or still consolidating under self-weight, and peat are the types of soil most likely to subside if the vibration is strong enough. Damage can be intensified if a hard stratum is present at moderate depth, as this reflects the vibrations back towards the foundations of the building at risk.

There are not many recorded cases of structural damage caused by traffic vibration, but there is at least a small risk of slight damage if the traffic is heavy or runs on an uneven surface, as anyone who has lived near a pothole in a busy road can confirm. Persistent traffic vibration can be a threat to inherently weak buildings (such as earth wall construction) and buildings that have been weakened through deterioration or previous structural movement.

Machinery can cause structural damage if it runs at a frequency that resonates with the building, or if it passes through such a frequency as it starts up or slows down. The greatest threat of damage is to the immediate structural supports – joists, beams, columns or walls. When new machinery is introduced, a structural engineer should calculate the building's natural frequency and check it against the information produced by the machinery supplier. Damage can be averted by careful siting of the machinery, by local stiffening of the structure, or (if practicable) by isolating the vibrations from their surroundings. If the vibrations pass through the building and its foundations, and into vulnerable ground, they can cause subsidence.

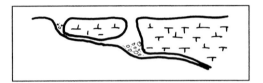

Figure 9.11
Swallow holes in soluble
rock.

Vibration damage is only progressive while the vibrations continue. Movement does not persist beyond that point unless the vibration has created a permanent weakness that makes the building sensitive to underlying problems.

Swallow holes

A swallow hole appears without warning as a surface depression. The sudden loss of support can be disastrous. Fortunately, disasters are much less common than near misses, which can be stabilized before they progress to greater misfortune.

Certain rocks and soils are soluble. The main examples in the UK are limestone, chalk and some chalky boulder clays. Ground water forms underground channels and wears away at local weaknesses such as fissures in the soluble rock, forming a cavern that grows until the roof above it becomes unstable and falls in, partially filling the cavern. The water continues to dissolve the rock. Eventually the disturbed material may migrate upwards as far as the surface (Figure 9.11).

If the soluble rock or soil is overlain by a different material, this too will fall into the solution cavern and migrate upwards; again, the end result is a depression at the surface (Figure 9.12). Loose sand is readily disturbed, and the initial surface depression is usually small in radius, so a swallow hole normally becomes apparent before it causes severe damage.

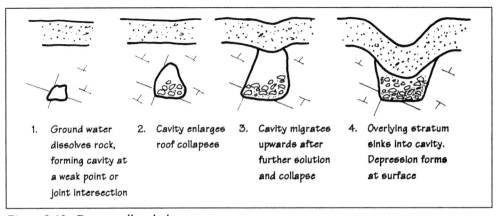

1. Ground water dissolves rock, forming cavity at a weak point or joint intersection
2. Cavity enlarges roof collapses
3. Cavity migrates upwards after further solution and collapse
4. Overlying stratum sinks into cavity. Depression forms at surface

Figure 9.12   Deep swallow holes.

On the other hand, if the overlying material is stronger than the soluble rock or soil, it will remain intact until the solution cavern is very large. In that case, unfortunately, when collapse eventually does occur, it can suddenly create a very wide depression.

Although the solution cavern may take many years to develop into a threat, the final breakthrough is often triggered by a sudden change in ground water level, or leakage, or by new loading above the cavern roof.

Shallow mining

Shallow mining shares many features with swallow holes: an underground cavern forms (this time man-made) and its roof collapses, leading to an upward migration of loose material, culminating in the appearance of a depression at the surface. Again, the triggering event is often an influx of water. Chalk workings, dry when the material was originally excavated, are particularly vulnerable to water, which may quickly enlarge underground tunnels by solution and erosion. The source of late water often turns out to be building development, introducing unsealed storm water drainage, or services that eventually leak if damaged or poorly maintained.

Mine shafts were capped or infilled when the mines were decommissioned, but some of the capping and filling was inept. The same potential exists for sudden loss of support.

Deep mining

Areas of deep mining are known, and movements are reasonably predictable. As a seam is worked, a wave of surface depression follows its progress (Figure 9.13, *overleaf*). Shortly after the start of this wave, the soil at the surface undergoes lateral strain – first tensile, then compressive. Most traditional buildings are unprepared for this, and the strain passes, little diminished, from soil through footings to superstructure. A wall aligned at right angles to the subsidence wave would suffer tensile strains typically of the order of 0.2% to 0.8%. This translates into 20 to 80mm of cracking in a 10m-long wall, or slightly less, allowing for the fact that masonry can sustain a small amount of tensile strain before fracturing. In the unfortunate event of the building being constructed above a geological fault or a heavily jointed rock stratum, the tensile strain may be accompanied by a vertical shear strain. The combination can be devastating.

Only minor movement is ever detected outside the angle of draw (Figure 9.13). This usually lies within the range of 25° to 35° to the vertical.

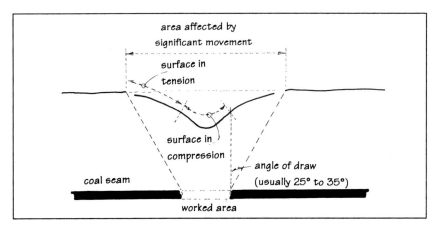

Figure 9.13   Mining subsidence.

If the rock strata above the worked seam dip sharply, the angle of draw and the shape of the subsidence wave are likely to be distorted.

When the compression strain has run its course, soon after the seam has been worked, the ground then stabilizes. Minor movement is sometimes detectable for up to two years after the start of the subsidence wave.

Modern tunnelling

Although tunnels cause no long-term ground movement, unless they fail, minor strains can be set up during construction. Where these might cause damage, surveys and monitoring should be carried out, as described earlier in this chapter (*Pumping and ground water lowering*).

Consolidation of made ground

Consolidation under self-weight should be distinguished from collapse compression, described earlier in this chapter, under *Inundation*.

Although consolidation of made ground is recognized as a hazard more often than it used to be, the proportion of buildings that have had to be sited on made ground has been steadily rising, so it continues to be a major and frequent cause of damage. It besets mainly new buildings. The movement follows the pattern shown in Figure 9.1 (*page 81*). Table 9.1 (*page 80*) gives a rough idea of typical times from deposition to 90% completion of consolidation, for various common types of made ground. Even allowing for substantial creep, few buildings more than 20 years old are at serious risk from continuing movement. Many older buildings still, of course, carry the battle scars.

An extinct or near-extinct consolidation does not normally need any fundamental attention, provided the building is still safe and has not developed precarious distortions. During the active years, structure and soil will have redistributed the permanent loading to limit the damage, and they will have fashioned an equilibrium between the weight of the building and a bearing pressure that is probably quite variable (Figure 4.2, *page 45*). Sometimes building professionals and their clients are persuaded by the visible building distortions and daunting ground conditions to seek a more orthodox equilibrium, usually at greater depth, and usually by underpinning. That is not often the best value for money. It may be merely closing the stable door after the horse has bolted. On the other hand, it must be recognized that the foundations will not enjoy the margin of safety expected of a well-designed new building, and they will be sensitive to any ground problem that might develop later. Escape of water (see earlier in this chapter) is probably the greatest menace to established buildings on made ground. This and other long-term risks can be managed by preventive maintenance (Chapter 23). Inundation of loose dry fill (also discussed earlier in this chapter) is less a frequent cause of damage, but more daunting when it occurs and less easy to manage or prevent.

Retaining wall failure

Figure 9.14 shows a building whose foundations rely on the stability of a retaining wall. If the wall were to fail, the effect on the building could be catastrophic. Sometimes, measurable movement (lateral or rotational) of the wall has to take place to mobilize the designed resistance of the soil on the downside. The wall might well be safe, but its less than rigid support can allow minor damage to the building. Any apparent movement of the wall, or minor structural damage to the building, should be investigated

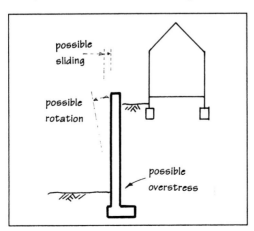

possible sliding

possible rotation

possible overstress

Figure 9.14
Retaining wall supporting building

sufficiently thoroughly to remove any doubt about whether the symptoms represent limited settlement or incipient failure.

Reducing external ground level

Figure 3.10 (*page 38*) showed the mechanism of failure in clay. Lowering the ground level alongside a building reduces overburden and risks failure. Non-clay soil can also be endangered by ground level reduction, especially if the soil is loose. Ground level lowering, to provide a parking area, footpath or access, is sometimes carried out without realising the potential consequences. It is not an everyday problem, but incidents have caused sudden and occasionally serious damage.

Trenching

A variation on the theme of lateral support loss is the service trench dug too close to a building's foundations (Figure 9.15). This has on occasion caused the collapse of a building.

Slope instability

The causes of slope instability were discussed briefly in Chapter 3. It can cause catastrophic damage to buildings. Although slopes sometimes advertise their incipient failure by tension cracks and initial surface movement, these symptoms do not necessarily become obvious in time for preventive action to be taken.

Most natural inland slopes are stable unless they are unintentionally weakened by steepening or loading or excavation. Excavation is the most common destabilizing agent. On one occasion, cutting out a wedge of sandy soil to form a small area for a barbecue was sufficient to start a slide, which exposed the piled foundations of a house twenty metres uphill.

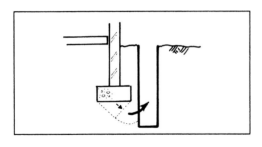

Figure 9.15
Failure caused by trenching.

Soil creep

Soil creep is the translational movement of a thin layer consisting of top soil and loose or weakened subsoil (Figure 9.16). It occurs because the restoring forces that prevent downhill movement fluctuate between adequate and inadequate. Low-grade chalk may be occasionally weakened by varying levels of ground water or by freeze–thaw cycles. Clay may be fissured by seasonal shrinkage, causing a slow downhill migration. Areas that were glaciated have been left, here and there, with unsorted material that once flowed when extremely wet. Present conditions are seldom as unfavourable, but a combination of ground water, temperature variation, vegetation growth and even burrowing animals may be sufficient to reinstate something like the original movement at a slow, stop–go pace.

Signs of creep are cracked or hummocky ground, and tree trunks and fences leaning downhill. These can also signify more deep-seated instability. If there is a more than insignificant likelihood of deep-seated movement and the consequences of it would be serious, a geotechnical engineer should be commissioned to provide a risk assessment.

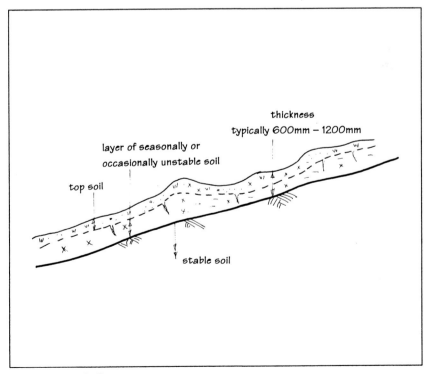

Figure 9.16 Soil creep.

Loading of adjacent ground

Bearing pressure is often depicted as diffusing downward at an angle of 45° (or flatter in weak soil). In practice, it forms contours of reducing pressure, whose pattern depends on the stiffness of the soil. If a new building is constructed adjacent to an existing one, it is likely that the pressure bowls will overlap (Figure 9.17), unless the new foundations are designed to transfer their load to the soil at a safe distance from their neighbours. Any significant rise in bearing pressure will make the older building undergo renewed, differential, settlement. The standard pattern (timing) of settlement (Figure 9.1, *page 80*) will be followed.

A soil whose stiffness increases linearly with depth, from zero at ground level, will not cause any pressure increase outside the building's perimeter. No soil matches such a mathematically exact model, but some are close enough to it for new buildings to be constructed with negligible effect on their older neighbours, so it would be wrong to jump to conclusions when diagnosing damage that appears to coincide with new construction. In some cases, the older building's new damage may instead be the legacy of temporary loss of restraint while the now perfectly innocent foundations were being constructed (Figure 9.15). This is the more common problem, and it is potentially the more severe. Diagnosis can be difficult.

Chemical attack

Foundations of traditional buildings may deteriorate in aggressive ground conditions or, in a few cases, because of deleterious materials in the foundation construction. Concrete and mortar are the materials most at risk. Attacks are rarely serious enough to cause damage to the building above ground level. When they do, specialist advice on the chemistry of deterioration is needed. Chapter 14 is a brief introduction to the better known types of chemical attack, some of which may occur above or below ground.

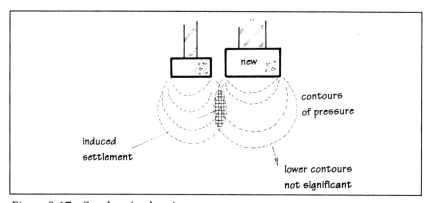

Figure 9.17 Overlapping bearing pressures.

Chapter Ten

Ground floor defects

Many ground-floor defects are mild and merely inconvenient, but some can lead to serious and progressive damage. This chapter describes the defects that commonly cause structural damage.

Ground-bearing concrete floors

It is difficult to tell by visual inspection whether floor damage originates in the covering, the screed, the slab, or the hardcore (sub-base) or soil (subgrade). Offices and houses are not challenging environments, as far as the ground floor is concerned; most of the problems that affect it arise from inherent construction errors (Figure 10.1, *overleaf*). The two most common are inadequate screed and hardcore settlement.

Inadequate screed

The concrete floor screed should be cement rich with a low water content. This is difficult to mix without a special screed mixer (not a normal concrete mixer). Inadequate mixing produces a variable material, in some areas weak and lean, in others too rich because the cement has not dispersed. The screed will also fail if it is laid too thin. (It should preferably be 65mm or more thick, if it is not bonded to the slab.) A poor screed may exhibit excessive cracking or indentation by loads. (Depending on the severity of any defects, and on practical considerations, screed can be either hardened by bonding agents, coated by a thin self-levelling skin or replaced.)

As a general rule, many small surface cracks are a sign of a screed problem. Fewer, larger cracks are more likely to indicate underlying structural movement. Other indicators would be needed for a firm diagnosis.

The slab can have mixing and curing faults, similar to the screed, but in this case quality is less critical. If it benefits from the protection of a screed, it is not often necessary to replace it solely on account of its own faults.

Hardcore settlement

During the 1960s, the single most common defect in new houses was the settlement of ground-floor slabs on their hardcore. Many builders underestimated the importance of stripping all vegetable soil, confirming a suitable natural formation, obtaining a supply of inert, well-graded hardcore and compacting it properly. As a result of education by the NHBC, the problem is now much less common, but it still crops up if one of those tasks is neglected. If the defect is restricted to poor compaction, prognosis is similar to a small-scale case of made ground consolidation (Table 9.1, *page 81*). Inclusions of organic material would prolong the consolidation.

It is worth noting that any organic material within the natural soil below the hardcore may add to the floor's settlement by its own consolidation, but very rarely does it become a serious progressive problem.

The importance of hardcore being well graded is still underestimated by a few builders. Sometimes, uniformly graded material is used, which is difficult to compact with standard mechanical equipment. However much effort is used, it remains below optimum density and is liable to consolidate during the first few years of use. This is rarely a severe problem with lightly loaded ground floors. It can be made worse if blinding sand is placed on top of the hardcore. The sand particles often fit neatly within the voids

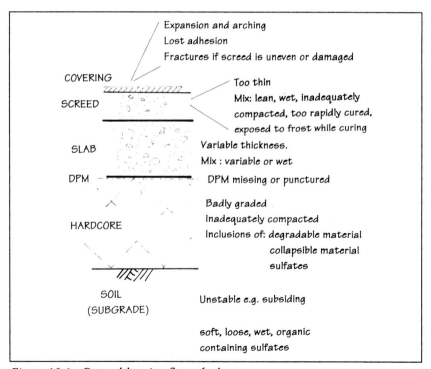

Figure 10.1 Ground-bearing floor: faults.

of the uniformly graded hardcore and, if encouraged to do so by normal events (pressure from loads, minor vibration, ground water), they increase total settlement by the original thickness of the blinding. An example of a popular uniformly graded hardcore material is 'gravel rejects', which have a particle size distribution within the coarse gravel–cobble region (Table 3.1, *page 22*). Gravel rejects do have a tendency to settle more than is normal, especially if blinded by finer material, although they usually stabilize within a few years and, at most, within a few centimetres.

Hardcore containing shale is prone to heave when the shale becomes wet. Fortunately, this is another problem that has diminished through education.

Hardcore containing material with sulfates, especially gypsum plaster, is also liable to swell and heave.

Subsidence and heave

Most of the subsidence and heave problems that damage foundations are likely to damage the ground floor at the same time. A near exception is clay–tree subsidence. Roots extend to the perimeter foundations and, at first, they damage only the outside walls supported on them. If the ground floor is independent of the walls, it will not be affected immediately. Obviously, if it is designed as suspended or incidentally built into the external walls, it will move in sympathy with them. If this sympathetic movement is firmly resisted by sub-base (hardcore) or subgrade (soil), it will probably crack close to the perimeter. If the offending vegetation lives and prospers, its roots may, before long, extend inwards below the building. They are likely to find the soil under the building to be a useful reservoir. Water content of soil is usually higher and more stable beneath a building than in the open, because a building (or any impermeable cover) gives protection from seasonal evaporation and, until the arrival of roots, transpiration. Thus, when roots are allowed to establish themselves under a building, they often find an abundant water supply, which they soon draw upon, causing the ground floor (if it is ground bearing) and internal walls to subside rapidly. Winter replenishment is very slow because, in the absence of indoor precipitation, water can only drain inwards slowly from the outside. With reference to Figure 9.5 (*page 85*) and the accompanying discussion, ground floor subsidence caused by vegetation in clay can quickly escalate from stage A to stage C.

Other ground-bearing floors

Brick-on-earth floors and timber-on-earth floors still survive, and most of them are at ease with the subsoil and need not be disturbed unless damp or deterioration have become manifest. They can be replaced with the same materials, assuming there are no deeper problems. Replacement would provide an opportunity to introduce a damp proof membrane, which should extend the life of the new materials. (But see Chapter 18, *Timber suspended floors*, regarding possible side effects.)

Suspended concrete floors

It is rare for an *in situ* concrete suspended ground floor to fail. When it does, the cause is most likely to be the result of careless fixing of the reinforcing bars. These should be close to the soffit but with adequate concrete cover. (The amount of cover depends on whether there is a DPM below the slab.) Occasionally, however, the cover is reduced or lost altogether. Whatever the cause, there is little option but to replace a failed slab.

Beam and block floors (Figure 10.2) also have a low failure rate, although more things can go wrong with them. Carbonation (Chapter 14), for instance, can weaken reinforcement by causing it to rust. If the floors comprise I or X-beams, and were constructed between 1930 and 1973, they may contain high alumina cement (HAC). The presence of HAC (discussed briefly in Chapter 14) raises doubts about the floor's capacity to support normal loading, which may be confirmed or resolved by either chemical tests or load testing.

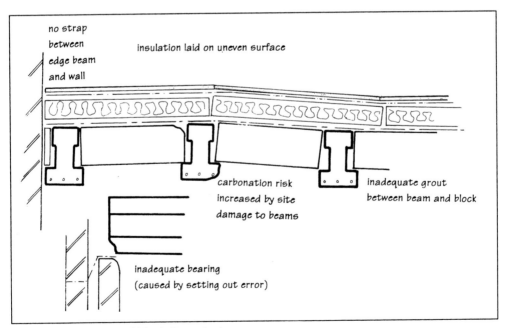

Figure 10.2 Beam and block floor: faults.

Any type of suspended floor, *in situ* or beam or plank, can be damaged by the agents of concrete deterioration, and it is difficult and expensive to carry out an investigation. If the floor sags (deflects) by more than 0.7% of its span, or if cracks show through the covering, a structural engineer should be instructed to carry out an appraisal.

Timber suspended floors

Timber floors can incubate problems for many years before the symptoms become obvious. Failure is the result of either deterioration or structural weakness. Causes of deterioration are discussed in Chapter 15 and listed in Figure 15.2 (*page 134*). Structural weakness may arise from:

- notching joists for services
- subsiding or deteriorating sleeper walls
- heavy loading introduced without checking floor capacity.

Concrete industrial floors

If industrial floors have to withstand a harsh environment, they may suffer from one or more of the following defects:

- settlement caused by heavy loading (particularly vehicle wheels or legs of stacking units)
- surface deterioration caused by wear
- surface deterioration caused by corrosive chemicals.

The following faults can permit damage:

- a weak slab surface (sometimes the top few millimetres consist of laitance produced during the finishing process, which acts like a brittle skin)
- a weak slab (if not designed for any particular loading, it may be too thin, too lean or of inadequate strength)
- reinforcement either insufficient in quantity or fixed out of position (the slab may then act as plain concrete and fail to distribute the load efficiently or to span across a variable subgrade)
- joints not working (if joints are at the wrong spacing or do not permit the intended movement, for example because dowels jam, then cracks and distortions may occur; joint edges may lip or spall)
- sub-bases poorly compacted
- subgrades loose or soft.

Chapter Eleven

Loss of equilibrium

No matter how distorted a building may be, if it is currently stable, it is in equilibrium. But equilibrium can be disturbed by unwise alteration, by serious accident, or by the introduction of heavy loading that creates, at least locally, forces beyond the building's capacity to distribute. This chapter discusses some of the ways traditional buildings can be hurt by loss of equilibrium.

Alteration

A building remains stable only as long as it is in both:

- external equilibrium – not overturning
- internal equilibrium – not breaking apart (Chapter 2).

Figure 11.1 shows an example of external equilibrium disturbed by the alteration of a medieval jettied building.

Less dangerous but more common assaults on timber-framed buildings include removal of cross ties and piercing of infills, in both cases usually to provide openings. More often than not, the structure accommodates

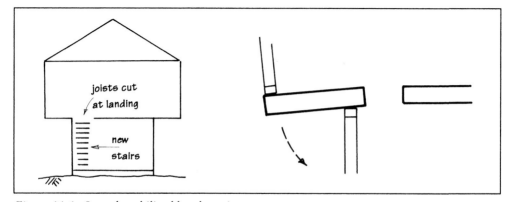

Figure 11.1   Jetty destabilized by alteration.

these alterations at the cost of some distortion and a loss of robustness, which may cause additional problems some years later.

Another common alteration – common to buildings of all ages – is the insertion of an opening in a load-bearing wall. It is a simple task, but it can lead to problems if the installation is not supervised carefully enough to ensure that everything above the opening is fully supported from the first cut to the last bit of pinning up. Sometimes the new beam is made strong enough to support its load, but not stiff enough. (A 15mm deflection may be acceptable above a 6m opening in a new building, because it accumulates as the dead load increases. The same deflection in an existing building is created suddenly and is more likely to cause damage, albeit cosmetic damage.) Figure 11.2 shows this and other potential problems.

Notching for services in timber joists and in masonry walls is often not considered to be structural alteration by those making the notches. However, unless it is carried out in accordance with calculations or safe rules, notching can have dangerous consequences. The rule with existing walls should be to avoid horizontal notching if at all possible (even though the building regulations permit limited notching in new walls).

Sometimes the new load paths enforced by structural alterations are more precarious than the alterers expected, if they depend on some detail that may or may not work. An example is the removal of a chimney below

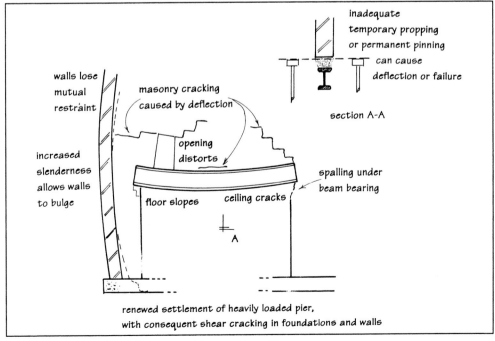

Figure 11.2 Possible defects caused by cutting walls.

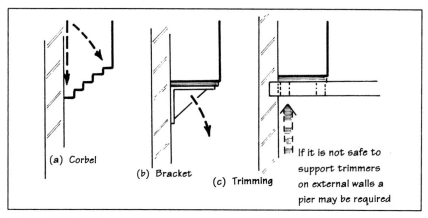

Figure 11.3   Chimney supports

the roof area, leaving a stump left standing in and above the roof space (Figure 11.3). The ability of brickwork to support loads on corbels (a) is much admired in partly demolished buildings, but it cannot be guaranteed. It relies on the shear and tensile capacity of brickwork, which is small and, for most design purposes, is best taken as zero. A bracket (b) is sometimes used instead of a corbel, but it imposes tension on the external brick wall, which again is risky. Positive support (c) is more reliable, provided that:

- its strength and stiffness are adequate
- the brickwork that ultimately supports the load is adequate
- the supporting joists or beams are protected from rot.

Accident

Impact provides an involuntary test of a building's robustness. Assuming that the accident does not remove a vital part of the structure, leaving no alternative load path, damage will usually be widespread but comfortingly slight if the building is robust, and especially if it is also resilient. Timber is the most resilient material. A building that is not robust will not be able to distribute the sudden force throughout its walls and floors. It will suffer more local but much more serious damage.

Accidents are liable to expose hidden weaknesses (see Figure 7.7, *page 70*). Similar comments apply to blast damage.

The force of a blast is much reduced if it is vented close to its source, by windows or a portion of the wall being blown out. In one case, the pitched roof of a hotel lifted and landed again with hardly any damage, and the relief it provided while briefly airborne meant that damage to the top storey was comparatively slight. It is unlikely that this fortunate outcome could be claimed as a design achievement, but there are occasions when accidents can and should be anticipated by risk assessment and precautions taken if appropriate (Chapter 23).

It is rare for blast or explosion to damage foundations, unless a crater is formed.

Overload

Overload is a less common cause of damage. Sometimes the full structural consequences of a change of use are not realized at the drawing board stage – perhaps not even communicated to the drawing board. One non-communicated amendment involved the last-minute provision of a greatly increased volume of cold water storage in the roof space of a commercial building. The *ad hoc* local trimming might have been adequate for a small roof; but the roof in this case was large, slender and not well restrained, and the additional load led to serious buckling.

Chapter Twelve

Movement

Buildings move continuously in response to normal variations in imposed loading, temperature and moisture. This is their means of maintaining equilibrium in the changing environment. This chapter describes some of the movements caused by moisture and temperature changes, and looks at some of the problems that arise when buildings are unable to cope efficiently with them.

Drying shrinkage

Table 12.1 summarizes typical drying shrinkage movements, which are irreversible. Table 12.2 summarizes cyclic moisture and thermal movements. Both sets of figures relate to free movement, but in real buildings there is usually a certain amount of restraint from adjoining walls and floors. Restraint reduces movement at the expense of incurring stress, which is the cause of 'shrinkage damage' or 'thermal damage'.

Table 12.1 Drying shrinkage

	Irreversible shrinkage $\times 10^{-2}$
Clay bricks*	nil
Calcium silicate bricks	0.01–0.04
Dense concrete blocks	0.02–0.06
Lightweight aggregate (autoclaved) blocks	0.02–0.06
Aerated (autoclaved) blocks	0.05–0.09

*Clay brickwork sometimes undergoes an initial irreversible expansion of 0.02×10^{-2} to 0.10×10^{-2}.

Table 12.2   Moisture and thermal movement

	Free moisture movement $\times 10^{-2}$	Free thermal movement $\times 10^{-6}$ per °C
Clay bricks*	0.02	5 – 8
Calcium silicate bricks	0.03 – 0.06	10 – 13
Dense concrete blocks	0.02 – 0.04	6 – 12
Lightweight aggregate (autoclaved) blocks	0.03 – 0.06	8 – 12
Aerated (autoclaved) blocks	0.02 – 0.03	8
Mortar	0.03 – 0.06	10 – 13
Granite	–	8 – 10
Limestone	0.004 – 0.013	3 – 8
Sandstone	0.025 – 0.070	7 – 10
Softwood	0.6 – 2.6 tangential 0.45 – 2.0 radial Nil longitudinal	4 – 6 with grain 30 – 70 across grain
Hardwood	0.8 – 4.0 tangential 0.5 – 2.5 radial Nil longitudinal	4 – 6 with grain 30 – 70 across grain

*Clay brickwork sometimes undergoes an initial irreversible expansion of 0.02×10^{-2} to 0.01×10^{-2}

Free movement
Figures for moisture movement assume, in the case of masonry, a change from dry to saturated and, in the case of timber, a change from 60% to 90% relative humidity.

Moisture: change of size = factor from above table × original size.

Thermal: change of size = factor from above table × original size × temperature change.

Free movement = moisture change of size + thermal change of size.

Actual movement = free movement minus restraint

Restraint reduces movement and introduces stress. It may act both in line with the movement and eccentrically, and it may be more effective in one direction than another.

Concrete blockwork

Severe shrinkage and damage can occur when materials are built in with extraordinarily high moisture contents, usually as a result of being saturated when stored without protection from the weather. Materials can also be saturated after they have been installed into the building but before it has been covered. Occasionally, concrete blocks are used straight from production, when they are green, and their potential shrinkage is higher than normal. The use of cement-rich mortars, such as 1:3 and 1:4, which are rarely necessary for structural reasons, also increases the shrinkage of walls made with concrete blocks.

Calcium silicate brickwork

In the case of calcium silicate (sand lime and flint lime) brickwork, the shape of panels between free ends, or between movement joints when these have been supplied and are working satisfactorily, also has an influence on the likelihood of cracking. Square panels are least likely to crack. The risk of cracking rises steeply when panels are either three times taller than their length or three times longer than their height. As with other types of masonry, openings may induce cracks.

Timber

The shrinkage of timber joists tends to be uneven. Abnormal shrinkage can create an uneven floor surface. If a partition is built off a timber beam or pair of joists, the latter's shrinkage, if severe, will leave it unsupported. Unless the partition can find some local points of support (say, at either end of the wall) and can span between them, it will suffer damage, typically developing diagonal and horizontal cracks. Openings are liable to distort. If a timber stud partition itself shrinks, it will create a gap at its head if it is non-load bearing, or distortions of supported members if it is load bearing. One common example of modern timber frame shrinkage is the inward tilting of window cills, as the internal skin of timber shortens while the external skin of brickwork remains the same shape. Local effects of shrinkage, in medieval timber frames, in the form of splits and twists, are usually benign once they have stabilized. Small-section timbers may suffer structural damage at joints, including joints secured by metal fastenings, which may become loose and ineffective (but usually can simply be retightened as part of normal maintenance).

General comments

Drying shrinkage is usually no more than a nuisance. It is a short-term problem, usually working its way out within a year or two. It can be regenerated if the building is rehabilitated after standing empty and becoming damp; or if there is a change of use, or even merely a change of owner, if the new owners prefer a higher internal temperature (or lower humidity). This can lead to damage that people find disconcerting because it appears unexpectedly.

However bad the shrinkage, or whenever it occurs, there is little option but to let it run its course and then to carry out whatever cosmetic repairs are desired. Structural repair is only occasionally needed.

Expansion

Moisture expansion

Whereas calcium silicate bricks and concrete blocks shrink irreversibly when they dry out, clay bricks expand after firing. The expansion is caused by the bricks absorbing atmospheric vapour. Table 12.1 notes the typical range of expansion, influenced by the type of brick and firing conditions. It can take several years for the movement to work its way out fully, but most of it is complete within a few weeks, usually before the bricks have been laid. For that reason, initial irreversible moisture expansion is not a common cause of damage except at times when storage of bricks after manufacture is brief, usually because demand temporarily exceeds supply.

Sulfate expansion

Brickwork is liable to sulfate attack if all of the following are true:

- The sulfate content of the bricks exceeds 0.5%.
- The mortar contains Ordinary Portland Cement (OPC).
- The wall stays wet for long periods.

The sulfates within the brick are dissolved by rain and migrate to the mortar, where they convert tricalcium aluminate in the cement to mainly calcium sulfo-aluminate. This chemical change is accompanied by expansion. Severe attack can lead to disintegration of the mortar. More common symptoms are horizontal mortar cracking and vertical expansion of the wall (typically 0.2%), caused by the thickening of the mortar beds, with consequential damage typical of restrained expansion. In lightweight buildings with cavity walls, the expanding outer skin has little restraint and can lift the roof – and sometimes the inner skin of brickwork – with it. This action creates horizontal cracks in the inner skin, often at or near roof plate level. If the expansion is partly or wholly restrained, the outer skin will usually bow outwards.

Render is damaged by sulfate expansion of the brickwork, and it will also be attacked by the sulfates if it contains Ordinary Portland Cement.

Sulfate expansion rarely becomes noticeable within two years of construction; this distinguishes it from irreversible moisture expansion, which appears quickly and is very seldom active enough to cause fresh damage after two years.

Calcium silicate bricks have too low a sulfate content to be vulnerable to sulfate attack, unless sulfates arrive in the cement from a different source.

Any substructure masonry made with OPC mortar would of course be open to attack by sulfates in the ground water, in the same way as concrete (Chapter 14).

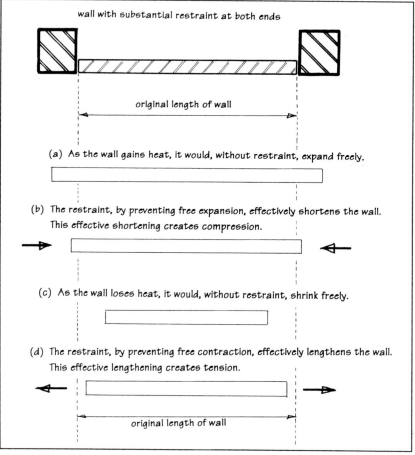

wall with substantial restraint at both ends

original length of wall

(a) As the wall gains heat, it would, without restraint, expand freely.

(b) The restraint, by preventing free expansion, effectively shortens the wall. This effective shortening creates compression.

(c) As the wall loses heat, it would, without restraint, shrink freely.

(d) The restraint, by preventing free contraction, effectively lengthens the wall. This effective lengthening creates tension.

original length of wall

Figure 12.1   Thermal loading.

Cyclic movement

Cyclic thermal movement

Figure 12.1 shows how temperature changes cause thermal loading in masonry, leading to compression during the expansion phase and tension during the contraction phase. If damage is caused, in the form of cracking, this nearly always occurs during contraction. Once formed, cracks open and close in response to continuing temperature changes. In the long term, the damage is slowly progressive, because when the cracks are open, dust from crack edges and wind-borne debris gets into them, so that they cannot close to the extent they did on the previous cycle. The next opening cycle starts from a new, partly open position, and the crack grows imperceptibly. After numerous cycles, noticeable widening can be observed to result from this 'ratchet effect'. Small buildings, such as detached houses, seldom exhibit severe 'ratchet effects'. Larger buildings and terraces may build up long-term distortions (Figure 12.3, *page 119*).

The simplest remedy, in theory, is to insert a movement joint so that the opening and closing continues with dignity. This is not always straightforward. The ideal movement joint is free from accidental restraint, a principle more easily applied to a brand-new building than an existing one. If inserting a joint is not practicable, repair should be carried out when the crack is near to its widest. This is likely to be during a cool day (but obviously it must not be freezing). The repair mortar should not be stronger than the existing mortar, because cracking will be less ugly if it recurs in the same place as before. By doing the work at the cool end of the temperature cycle, future cycles will be biased towards compression of the masonry, as the temperature will usually be higher than it was when the repair took place. Maximum tension will thereby (in most cases) be less than it was previously. It may even be reduced to below critical tensile strength, the point at which masonry cracks; in this case damage will not recur (Figure 12.2, *overleaf*). Although this sort of repair is a crude alternative to a joint, it is much cheaper, so it may be an acceptable first resort. It need not preclude a joint being installed later on if the cracking returns.

Cyclic moisture movement

All masonry undergoes cyclic expansion and contraction in response to changes in moisture content (Table 12.2, *page 113*), but this seldom causes structural damage. To some extent, the moisture movement modifies thermal movement. On a hot day, for example, a masonry wall will expand thermally, but any evaporation caused by the heat would encourage a moisture contraction (usually smaller).

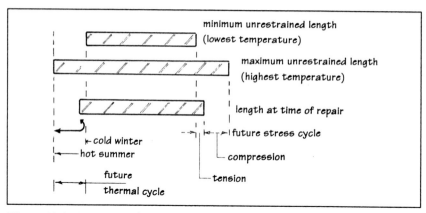

Figure 12.2   Repairing thermal cracking.

Critical changes of shape

Cyclic movement and creep both change the shape of buildings. Occasionally the change, although seemingly small, can be critical. Some examples are shown in Figure 12.3. These show:

- a reduction of the length of a beam bearing, leading to over-loading and failure (a)
- lateral movement of a large timber roof, leading to buckling or side-sway (b)
- similarly, movement can cause long stud walls to rack, changing panel shapes from rectangles to parallelograms
- rupturing of junctions caused by repetitive movement (such as thermal expansion), leading to loss of robustness (section A–A)
- lean of end walls of long buildings or terraces, caused by thermal movement of the front and rear walls (c)
- creep of heavily loaded timber, leading to load redistribution or rupture of junctions (d).

Examples (a) and (b) have been known to lead to collapse. The others seldom cause more than unwelcome distortions, or repetitive damage. Nonetheless, they should always be investigated and, if in doubt, monitored so that any necessary repairs can be proposed to maintain safety, economy or simply appearance.

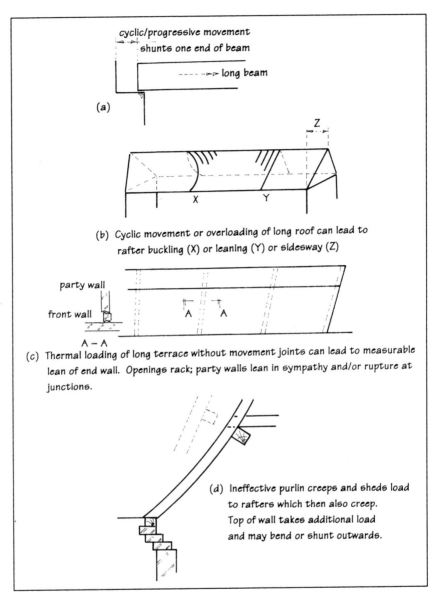

cyclic/progressive movement
shunts one end of beam

----▷ long beam

(a)

Z

X   Y

(b) Cyclic movement or overloading of long roof can lead to
rafter buckling (X) or leaning (Y) or sidesway (Z)

party wall

front wall

A   A

A – A

(c) Thermal loading of long terrace without movement joints can lead to measurable
lean of end wall. Openings rack; party walls lean in sympathy and/or rupture at
junctions.

(d) Ineffective purlin creeps and sheds load
to rafters which then also creep.
Top of wall takes additional load
and may bend or shunt outwards.

Figure 12.3   Cumulative effect of small movements.

Chapter Thirteen

Fire

Fires can damage a building by creating abnormal stresses during heating and fire fighting, by weakening structural materials and by creating enough debris to overload floors. This chapter outlines these problems and briefly describes the behaviour of traditional materials in fire.

During fire, the building is challenged by a series of events:

- Heat causes expansion.
- Heat drives out moisture, causing rapid shrinkage of timber and spalling of concrete and masonry.
- The properties of materials change for the worse, as their temperature rises.
- Fire fighting water causes thermal shock and saturation.
- Collapse of burned material and parts damaged by fire fighting creates debris that the floor, probably weakened, may fail to support.

Assuming the building survives the fire, it faces more trouble:

- The cooling structure contracts.
- Damage may occur, or recur, as thermal loads change.
- Until protection is in place, rain, wind and frost can cause damage.
- Drying out causes shrinkage and may instigate rot.

Non-structural problems, especially smoke damage and damage to services, add to the building's troubles and hamper investigation.

Not every fire creates every problem listed above, but the building should be examined for symptoms of them all before repairs are specified. Fires often remain local, through early quenching, through failure to grow because of resistance to flame spread, or through venting, lack of oxygen or lack of fuel. In these cases, it is often possible to find the seat of the fire, and its radius of significant damage. Radius is a misleading term, as in most cases heat travels upwards and then laterally at the highest level it can reach, often causing heat damage beyond the spread of flames.

The structural condition of a fire-damaged building depends on:

- the various stresses it suffered as a result of thermal and debris loading
- the permanent distortions left after expansion and contraction
- the residual strength of any structural materials that have endured high temperatures.

Thermal loading causes stress by expanding the building as a whole, so that perimeter walls are pushed outwards. In addition, individual walls may be fiercely heated on one side, and later cooled differentially by fire fighting. Heating encourages expansion, which does not directly cause stress; but the restraint it inevitably meets (Chapter 12) will certainly cause stress to the heated element and, very likely, to the restraining element. Table 12.1 (*page 112*) includes values for free thermal movement, from which it can be seen that masonry buildings will typically attempt to expand by about 0.1% per 100°C. Resistance to such movement can damage junctions and weak points. Where dissimilar materials meet, differential expansion may exaggerate the damage. Fire fighting water will cause rapid differential contraction, possibly followed by saturation, which may cause further movement; and, again, damage is likely to occur at junctions and weak points.

Permanent distortions should be measured to check equilibrium and to estimate stresses locked in by movement and eccentricity. If everything has returned to its normal shape, the main legacy of thermal loading is likely to be cracking that formed at maximum stress. There may also have been a loss of section size through spalling or charring. Members should be supported (propped or shored) if they are found to have been significantly weakened by the fire or to be supporting a heavier load from saturated debris than they might have carried during normal use. Walls left exposed and standing as ruined stumps should be carefully shored or safely demolished.

With the exception of timber, traditional building materials are permanently weakened by high temperature. Knowing the temperature contours would provide a very good first clue to the extent of weakening. This is not easy. There are formulae, based on furnace tests, for the gas temperature during a standard fire. Temperatures early in the standard fire depend on type of fuel, but in all cases there is a very steep rise in the first few minutes followed by a slower increase. Real fires may grow more slowly from ignition than the standard might predict, and at first they often appear to be innocuous and easy to control. But given enough fuel and an abundant oxygen supply, they will reach flashover, when the temperature is likely to rise even faster than the formula would predict, very quickly reaching the plateau temperature of a fully developed fire. Heat is transferred from the gas (atmosphere) to individual walls, beams and columns by radiation and convection, and it continues by conduction through solid material. Real fires decay, either naturally or by the efforts of fire fighters, after reaching whatever temperature they can (fully developed or otherwise) in the time available. An estimate of the temperature reached may be made following discussion with the fire brigade and inspection of non-structural fixtures, decorations and debris. Table 13.1 lists a few indicators of temperature. Observation on distortions, damage and the condition of materials caught in the fire are of great interest, although interpretation requires specialist knowledge.

Table 13.1   Fire temperature indicators

Temperature (°C)	Material behaviour
120	Polystyrene collapses.
150	Polythene melts. Plastic-encased equipment softens.
200–250	Plastic-encased equipment progressively distorts and collapses. Glass begins to soften. Wood and paper darken.
250	Soldered plumbing joints come adrift. Plastic light shades melt.
300	Siliceous concrete turns pale pink.
350	Lead melts.
400	Aluminium softens.
650	Aluminium melts.
800	Glass flows.
800–1000	Brass fittings melt.
900	Limestone and lime turn to quicklime. Siliceous concrete turns buff.
950	Silver melts.
1100	Copper melts. Radiators collapse.

Traditional materials and fire

The following paragraphs briefly describe the behaviour of various traditional materials.

Clay brickwork

Clay bricks are manufactured by firing, and clay brickwork performs reasonably well in fires, but it loses strength at high temperatures. This loss is greater if the wall is in bending as well as compression; failure, if it comes to that, occurs sooner. It is impossible to give a formula for degree of weakening, because it is highly sensitive to actual conditions during the fire and the fire fighting. In fact, loss of strength can vary markedly over short distances in the same wall. The mortar often weakens more than the bricks. Lime mortar turns to quicklime, and has little residual strength, if it reaches 900°C. After fire, brickwork should be examined for mortar damage, which is likely to be the first symptom of weakening, whether or not it contains lime. It should also be checked for spalling or horizontal cracking, which may be a sign of bending stress caused when one face reaches a higher temperature than the other.

Calcium silicate brickwork

Loss of strength of calcium silicate bricks is much greater than for most clay bricks, and greater levels of damage can be expected.

Limestone

This loses carbon dioxide above 500°C and turns to quicklime at 900°C. These mark the points of significant weakening and destruction.

Sandstone

Spalling can be serious above 100°C.

Concrete

For concrete or mortar containing siliceous aggregate, there is typically a 25% loss of strength at 400°C and a 75% loss at 600°C. Concrete or mortar containing limestone aggregate may deteriorate more quickly. Loss of strength is accompanied by changes in colour. For concrete containing siliceous aggregate, the concrete surface changes to a pale pink at 300°C, turning to whitish grey at 600°C and buff at 900°C. Limestone aggregate turns light pink at 200°C and dull grey at 400°C. These colour indicators are typical, but not inevitable, because minor changes in constituents have unpredictable effects. In some cases, there is a period fairly early in the fire when concrete surfaces spall explosively. Non-violent spalling occurs later, typically exposing coarse aggregate or reinforcement. If reinforcement is exposed, structural failure may occur rapidly. Steel reinforcement loses half its strength at 550°C; prestressing steel, often present in precast concrete floors, loses half its strength at 400°C. These figures are rough; the type of cement and aggregate all influence actual behaviour.

Timber

Charring of timber beams occurs at the rate of roughly 40mm per hour and columns at 50mm per hour. This happens roughly 25% more slowly in the case of oak. Slight rounding occurs at corners. Remaining timber is not weaker than it was before the fire, provided that metal bolts or flitches have not conducted heat into its interior.

Earth walls

The composition of earth walls varies, and so does performance, but there is a high risk of shattering during the fire, particularly if chalk is a constituent, and of sudden loss of strength if drenched by fire fighting water or left flooded afterwards.

If there is doubt about the residual strength of a member after a fire, core samples should be taken and tested.

Weather

It may take some time for repairs to get underway, and it is essential to protect the building from the weather. Waterproofing must have a stiff support and must drain water efficiently off the building. The interior should be dried as quickly as possible. During the drying-out period, timber should be monitored, as there is a risk that it will start to rot; it will certainly shrink. Covered timber, particularly timber within masonry, such as wall plates and bonding timber, should be exposed and ventilated to reduce the risk of dry rot.

As an alternative to these temporary measures, which add costs without adding value, opportunity might be taken to replace vulnerable timber, provided there is no objection on conservation grounds.

Chapter Fourteen

Chemical attack
on concrete

Ordinary Portland Cement is an ingredient in most concrete and post-1930 mortar and render. It is the building material most vulnerable to chemical attack. This chapter introduces the most common problems associated with it. Concrete chemistry is a specialism in its own right, and an expert should be engaged if a serious problem is suspected.

High alumina cement

In the third quarter of the twentieth century, many floors were made from precast concrete beams containing high alumina cement (HAC). This cement encouraged very high early strength, compared with Ordinary Portland Cement (OPC), so it was popular with manufacturers and their clients because the products could be moved out of casting moulds quickly. Early problems with HAC concrete were attributed to using too high a water–cement ratio in the mix, aggravated by imperfect curing conditions. It also became apparent that the finished product was less than ideal in warm, humid conditions. A few collapses in the 1970s prompted research, which concluded that, even with the best quality control, HAC concrete loses strength over time, owing to a change in the structure of the hydrated cement. The process is referred to as 'conversion'. Eventually, concrete strength may be only half its early peak. The rate of conversion is extremely variable but, since manufacture of the product ceased before 1980 in the UK, most HAC concrete is now highly converted and unlikely to become much weaker.

Many buildings were investigated and strengthened during the 1980s. Any building that has escaped testing could be a problem – even if its strength is no longer declining – because, although it may not have previously supported the full design-imposed load, it may have to do so in future. Without calculations or tests, it should not be assumed to be safe. Doubts about its strength will increase if the concrete has faults in addition to those associated with HAC. Long-term weakening is exaggerated if the concrete is in alkaline conditions or is persistently wet, or both. For example, a roof beam constructed in HAC concrete and supporting a screed containing OPC (which will be alkaline) may suffer additional deterioration if water is allowed to soak downwards, leaching OPC into the structural concrete. For this reason, roofs are more likely than internal floors to fail.

HAC concrete beams were prestressed using small diameter wires close to the outside face; in these conditions rusting is possible, which also increases the risk of failure.

HAC concrete tends to turn brown, but this is not a reliable way to identify it. Tests are available for recognizing the presence of HAC, the current strength of the concrete and the potential for its prestressing wires to rust.

Some HAC has been used in foundations because of its beneficial performance in certain aggressive ground conditions. Nearly all of it was in mass concrete, and the loss of strength has not prevented the foundations from performing their function.

Carbonation

Concrete is alkaline, which is essential to the durability of its reinforcement. In acid conditions, with a supply of water and oxygen, the reinforcement rusts, losing strength and causing damage to the surrounding concrete as it expands. The most common agent for rusting is carbonation. This is a natural process, driven by the slow penetration of carbon dioxide into the surface of the concrete and causing it to change from alkaline to acid. Carbonation is inevitable, but harmless on its own. The harm comes only if oxygen and water can then reach reinforcement in carbonated concrete.

Chloride attack

Another agent for rusting is calcium chloride. This may be present within the concrete, or it may penetrate from outside. Calcium chloride has been used as an additive to accelerate the setting of concrete. Its use declined in the early 1970s, when its potential for corrosion became generally known, but it did not altogether cease. Sodium chloride can originate from poorly washed marine aggregate, from salt spray in coastal areas and from de-icing salt applied to roads and parking surfaces. Chloride attack does not need carbonation to start it off, but the combination of carbonation and chloride attack will exaggerate the final damage.

Reinforcement rusting

The first sign of reinforcement rusting is usually staining of the concrete surface parallel to the reinforcing bars. As rusting progresses, the concrete spalls away to expose the reinforcement. If allowed to run its course, the defect will eventually weaken the reinforced concrete member to the extent that it will need to be demolished and replaced. If the early signs are recognized, a less drastic remedy can be applied.

Concrete is most at risk if it is weak, porous, contains large cracks for any reason, and provides less than the recommended cover to its reinforcement.

Sulfate attack

Sulfates dissolved in groundwater can attack concrete in foundations. There are two main types of sulfate attack:

- *ettringite*, where the calcium aluminate in the concrete expands when attacked by sulfates, leading to expansion and disruption of the concrete over a number of years
- *thaumasite*, where the sulfates attack calcium silicate, the main strength-giving chemical within the concrete.

Sulfates are present in groundwater dissolved out of overconsolidated clays, such as London clay, Oxford clay, Kimmeridge clay, lias and Keuper marl. They may also be found in peaty moorland water and on some contaminated or polluted sites.

Mass concrete foundations often have sufficient reserves of strength to withstand modest sulfate attack. If not, the only reliable treatment is to provide a barrier against invasion by soluble sulfates or, if the attack is advanced, to replace the damaged concrete. Reinforced concrete may be at greater risk if the attack leads to reinforcement rusting.

Concrete is most at risk from sulfate attack if it is weak and porous, and its surface area to volume ratio is high. In the case of thaumasite attack, limestone aggregate is the trigger.

Mortar below ground is equally vulnerable, unless made with one of the cements that resist sulfates. Sulfates can also originate from acid rain (Chapter 15) and flue gases, as well as from the bricks themselves, causing possible disintegration of the mortar.

Alkali–aggregate reaction

Another form of degradation is generated by an internal reaction between the constituents of the concrete. This is alkali–aggregate reaction, often referred to as alkali–silica reaction (ASR), the most common variation. Certain types of silica contain minerals that react with the alkalis in the concrete mix, leading to swelling and disruption. Deterioration is usually slow, and often insufficient to cause failure, but there is no remedy; in severe cases there is no alternative but to replace the elements affected. The onset of ASR can sometime be recognized by the small-scale random cracking ('map cracking') on the surface of the concrete.

The above discussion hardly scratches the surface of concrete durability, which in any case is not a major problem in traditional buildings. When it is suspected, an expert in the subject should be consulted.

Blockwork containing 'mundic'

In the second quarter of the twentieth century, many small buildings were built of locally made concrete blocks. A proportion of these blocks contained iron pyrites ('mundic') and other deleterious minerals within the concrete aggregate. The buildings were nearly always cement rendered. If water and oxygen can get to the blockwork, it will expand and degrade. The first visual signs are usually cracking and loosening of the render, which may appear as a random 'map' pattern or may follow lines of bed and perpend joints. Unfortunately, by the time the damage is usually recognized, it may not be possible to arrest the degradation other than by replacing the affected walls.

Mundic problems crop up mainly in southwest England, and local surveyors are able to advise on the risk of degradation. They are aware of the regional distribution of concrete containing deleterious aggregate, as well as the distribution in time. (A much smaller proportion of buildings constructed after 1950 are at risk.) They may need to take core samples, have them analysed and interpret the results. Although it is difficult to cure the problem once it has a hold, it can be prevented by keeping moisture out of the blockwork. Rainwater should be kept at bay by ensuring that the render is well maintained and regularly painted with a vapour-permeable paint; alternatively, cladding would provide a more positive barrier. An interior threat to mundic comes from condensation, so a good standard of interior heating and ventilation should be maintained.

Chapter Fifteen

Deterioration

Deterioration aids and abets other causes of structural damage by weakening members and their connections until they can no longer cope with everyday demands. Most of the time, deterioration is slow and incremental, often secret, but it can accelerate disconcertingly when vital protection is lost through lack of maintenance or unattended structural damage. The speed of deterioration usually decides the urgency of any intervention. Even when it is not contributing directly to the damage under consideration, it may have to be acknowledged as an influence on the choice of repair. This chapter introduces the most important agents of deterioration.

It is not useful to list material life expectancies, because so much depends on factors specific to each building, such as design, site exposure and quality of maintenance. Practising surveyors are the best people to make individual predictions, but they find it difficult to supply average figures. In a 1992 survey (*Life Expectations of Building Components*, published by RICS), the life of brickwork, for example, was given a median value of 100 years; but there was a very wide disparity of opinion between the individual surveyors who supplied opinions. Owners have more consistent views. They assume their buildings will last indefinitely, given normal maintenance. Normal maintenance may mean occasionally replacing damaged elements, but few owners expect to be advised to budget for rebuilding entire walls when the 100 years are up. How long does brickwork really last?

In fact, all traditional materials are very long lasting and can certainly provide service for several generations, but only if the original construction was of good quality and the various agents of deterioration are kept at bay by regular and intelligent maintenance.

Normal ageing

Ageing results in loss of mass or strength; it is brought about by one of several natural agents, including:

- atmosphere (chemical change, especially in polluted areas)
- wear and tear (impact from wind-borne debris; dynamic forces during use)
- rain (impact, solution, dispersal of dirt)
- inherent damp (condensation and ground water)
- frost (freeze/thaw cycles).

Loss of protection

Loss of protection is the greatest accelerator of deterioration. The most dramatic examples involve rain penetration, particularly from gutters, leading to timber rot. Exposed masonry also goes downhill if it loses its protection against prolonged saturation, which happens if copings are removed or damaged. Clogged ventilation can encourage dry rot in cellars and below-ground floors, and even in roofs. Disused chimneys may become damaged, and may cause damage to their surroundings, if hygroscopic salts are allowed to migrate inwards as a result of normal upward ventilation being frustrated.

External attack

One of the most dangerous agents of deterioration is man. The most common form of attack is inappropriate maintenance. The best-known example is brickwork pointing (Figure 15.6, *page 142*). Other zoological agents include:

- masonry bees (destroying mortar)
- rodents (chewing into earth walls)
- vermin in general (by their living and dying).

Botanical agents include:

- creepers (damaging the faces of walls by their anchorage)
- algae and lichen (breaking down stone and delaminating tiles and shingles)
- larger vegetation (when allowed to take hold in neglected valley gutters and flat roofs, damaging coverings and structural materials).

Fungi and tunnelling insects play a major role (see below, under *Timber*).

Corrosion

Rusting of ferrous metal is the most frequently observed example of corrosion, but all metals corrode. The science of corrosion is complex; we will mention only some basic principles. The two most important types of metal corrosion in buildings are: oxidation and electrolytic action.

Oxidation

Oxidation is the conversion of metal to metal oxide. The process is fuelled by the ample supply of oxygen from the atmosphere, and it is encouraged by water vapour. Below a relative humidity of 70%, the rate of oxidation is often low to negligible. As oxidation proceeds, a film of metal oxide forms on the surface of the metal. This chemical change is in most cases accompanied by a volume change. In other words, the oxide occupies a different volume from the parent metal. Ferrous oxide, for example, has a much larger volume than iron or steel. The rust, generated by embedded bolts, brackets and beam ends, expands and creates bursting forces in the surrounding masonry, whose effects are often more spectacular than the rust itself. With some metals, the oxide volume is smaller than the metal, and does not cause bursting.

The volume difference between metal and oxide is usually enough to cause disruption between them, so that they part company, presenting a fresh metal surface to the atmosphere, and allowing oxidation to continue. Aluminium, chromium and nickel are exceptions. Their oxides are not significantly different in volume from the metal and therefore the initial film remains intact and adheres to the metal, inhibiting further corrosion. Stainless steel is an alloy that produces a corrosion-resistant skin of oxide on its surface. Stainless steel is, in fact, a generic term for various corrosion-resistant steels containing at least 10.5% chromium as well as lesser quantities of other alloying elements. The austenitic range (from which most building stainless steel is manufactured) has the highest resistance to corrosion of all the stainless steels.

Electrolytic action

If two areas with electric potential are connected by an electrolyte, current flows from the lower potential to the higher, accompanied by a transfer of metal. The two areas may be metals of differing potential. They can, in some circumstances, be the same type of metal, and even the same piece of metal; if the conditions are right, a nail may corrode by, for example, transferring metal from its shaft to its head.

The electrolyte can be water, acid, a naturally occurring salt in solution, or a chemical introduced by treatment, such as fire retardant. Table 15.1 is a typical electrolytic table, yielding the following information. Where any two metals are present and connected by an electrolyte, there is a risk that the metal that is lower in the table (and therefore has the lower potential) will act as a sacrificial anode and migrate towards the higher (the cathode). The lowest metals are often called base metals; the higher ones are referred to as noble. As a general rule, the further apart the two metals are in the table, the higher the rate of corrosion, other factors being equal. (One factor that is never equal is the make-up of the electrolyte.) Metals that are close together in the table are usually safe in dry conditions. Lead and tin, which are adjacent, are safe in nearly all conditions. The order within the table varies to a limited extent, depending on the electrolyte, so Table 15.1 should be taken as typical, but not universal. It is safe to use different metals in close proximity in the building, even in adverse conditions, provided each is isolated from electrolytic contact with any others by washers, bushes and other isolating membranes.

Table 15.1   Electrolytic series

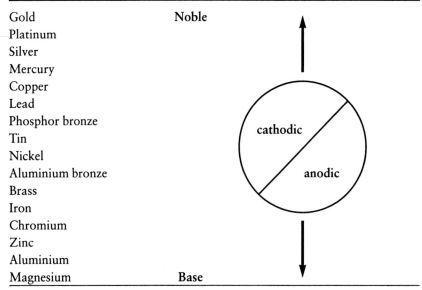

Gold	Noble
Platinum	
Silver	
Mercury	
Copper	
Lead	
Phosphor bronze	
Tin	
Nickel	
Aluminium bronze	
Brass	
Iron	
Chromium	
Zinc	
Aluminium	
Magnesium	Base

Note: The order shown is typical, but is influenced by the electrolyte. Minor variations occur in practice.

In some cases, preservative chemicals undergo short-term reactions that promote electrolytic action. Unless the installation of metal fastenings is delayed until these reactions have run their course, the risk of corrosion is considerably heightened.

TIMBER

Timber's enemies are water and tunnelling beetles. In combination the two are deadly.

Moisture

Figure 15.1 shows how the environment affects timber moisture content and vice versa. There are several implications to be drawn; the most obvious is that timber with a moisture content of 20% or less is safe from rot. Less obviously, timber that is being dried out after having been saturated (the building may have been neglected and left open to the weather, for example, or may be recovering from fire quenching) will be facing, temporarily at least, an increased risk of rot. This would have to be anticipated by preventive action; alternatively, it should be carefully monitored.

Figure 15.1 contains some simplifications. It does not show the effect of temperature or ventilation. Wet rot prefers a comfortable internal temperature to the colder and more variable external UK conditions, another

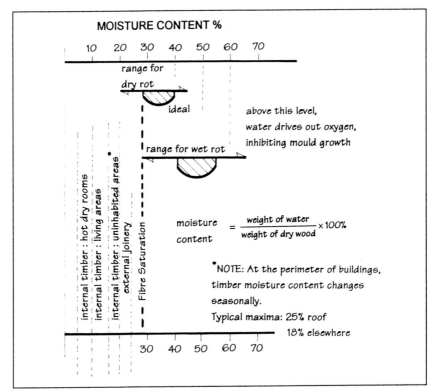

Figure 15.1   Moisture content and rot.

reason why the risk of it damaging timber may increase as a building is being dried out. Even a change of occupancy, if it involves a favourable shift in temperature, may sometimes be enough to instigate or rejuvenate rot. Nevertheless, wet rot will die out all together if the moisture content falls permanently below the threshold for germination, and its demise

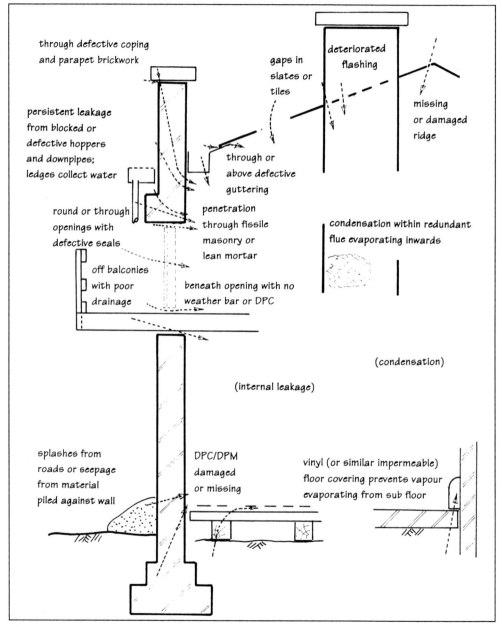

Figure 15.2   Sources of damp.

happens within weeks or months. Dry rot has greater persistence, because it has the ability to conduct water from one area of timber to another, and thus maintain a toehold (or strand-hold) on comparatively dry material for some time. This is why pains are taken to eradicate all signs of dry rot from the building. Not even dry rot, however, can thrive indefinitely on timber with a moisture content permanently below 20%. It also prefers a stable temperature, and is seldom found in roofs, where variations are not far removed from outside temperatures.

Internal wet rot need never happen in a well-maintained building; when it does occur, it demands urgent attention because it always affects, or threatens, structural members. Rot destroys the fibres of timber, making it valueless as a structural element; but, before it reaches that stage, the increased moisture reduces its strength and stiffness, compromising its load-carrying capacity.

Figure 15.2 shows some of the areas where damp most easily penetrates and rot is most likely to take hold. Many of these areas are difficult to inspect. Figure 15.3 is a reminder that interstitial condensation can allow rot to develop in areas that are impossible to inspect without opening up. Timber and timber products are the materials most vulnerable to damage by interstitial condensation.

There are several species of rot. In UK conditions, however, it is sufficient for nearly all purposes to recognize two groups of wet rot, white and brown, and one species of dry rot, *Serpula lacrymans*. Identification should be entrusted to an experienced surveyor.

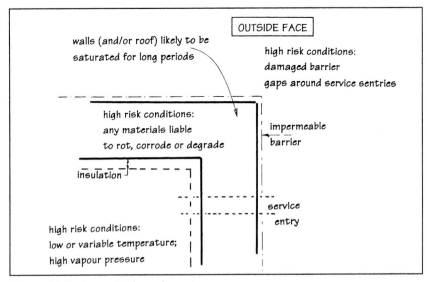

Figure 15.3 Interstitial condensation.

Table 15.2   Wood boring insects

Species of insect	Preferred wood	Characteristics	Life cycle (years)	Treatment
Common furniture beetle 'woodworm' (most common wood borer in UK)	Softwoods and European hardwoods. Sapwood and rotten heartwood. Moisture content must be 12% minimum.	1–2mm dia tunnels predominantly in direction of grain. 1–2mm dia exit holes.	3 plus	Insecticide to face (if active).
House longhorn beetle	Sapwood of softwoods. Most prevalent in roofs. Only common in Surrey and adjacent counties. Evidence of new infestation must be passed to BRE.	6mm to 10mm dia oval tunnels join to form large cavities. 6mm to 10mm exit holes have ragged edges (bore dust is cream coloured and the size of coarse sand).	3–11	Insecticide to face if attack is slight. Severely damaged timber should be burned.
Deathwatch beetle	Sapwood and heartwood of hardwoods (especially oak) but only if decay has started. Occasionally affects softwood in contact with or near to infected hardwood.	3mm dia tunnels mainly in direction of grain but very extensive, creating large cavities in severe attacks. 3mm dia exit holes.	3–4 typical; 12 max.	Insecticide injected (pressure or gravity) or smoke. Very difficult to treat cavities.
Powderpost beetle (not common in structural timber)	Sapwood of hardwoods (especially oak and elm).	1–2mm dia tunnels predominantly in direction of grain. 1–2mm dia exit holes.	1	Insecticide to face (if active).
Ptilinus beetle (rare)	Some hardwoods (beech, elm, maple, hornbeam).	1–2mm dia tunnels predominantly in direction of grain. 1–2mm dia exit holes.	2	Insecticide to face (if active).
Wood boring weevils	Any, provided decay has started.	Ragged 1mm dia exit holes.	1	Eliminate rot, dry out. (Replace decayed timber).
Wharf borer beetle	Any, provided decay has started.	6mm dia oval exit holes.		Eliminate rot, dry out. (Replace decayed timber).
Tenebrionid beetle	Any rotten timber.	Irregular large (10mm or so) exit holes.		Eliminate rot, dry out. (Replace decayed timber).
Stag beetle	Any rotten timber especially oak and elm.	Irregular, very large (up to 20mm) exit holes.		Eliminate rot, dry out. (Replace decayed timber).
Leafcutter bees, solitary wasps (rare indoors)	Decayed wood. Normally outdoors or near external face.	Circular exit holes 7mm dia.		Eliminate rot, dry out. (Replace decayed timber).

Chemical attack

Wood can be attacked by acid or alkali, and may at first look as though it is harbouring rot; but in most cases the effect is superficial and does not cause serious weakness. A common type of chemical attack is sulfur dioxide from soot. Metal fasteners are nearly always at greater risk than the surrounding timber.

As mentioned earlier in this chapter, metals may corrode if connected by an electrolyte. Timber can act as such if it is damp, especially if it:

- is naturally acidic (the heartwoods of oak, Douglas fir and sweet chestnut are more acidic than most)
- is naturally permeable
- contains salts, preservatives or flame retardants.

The products of corrosion will usually attack the wood in the vicinity of the cathode, which may, if it is a fastener, lose its grip. This is the cause of 'nail sickness'. Without intervention, the fastener will become useless, either through its own corrosion or through damage to the surrounding timber.

Insects

Wood is a food source for certain insects. With some species (termites, for example), it is the adults that attack the wood. In others, the damage is done during the larval stage. In the UK, the latter is almost exclusively the case. The life cycle starts when eggs are laid on or just below the timber surface. Emerging larvae tunnel into the wood and feed there, often for long periods, until the time comes to form pupae in chambers just below the surface. From there, the emerging adults bore through to the surface, leaving the characteristic circular or oval exit holes that advertise infestation.

Table 15.2 lists the most common species of beetle in the UK. Only five ever need to be treated with insecticide to prevent further attack. The others rely on the timber being wet, and the attack stops soon after the timber is dried out. When infestation is discovered, an experienced surveyor should be asked to identify the species (there may be more than one) and to advise on whether treatment is necessary.

Further points

Treatment and structural assessment of decayed timber are briefly discussed in Chapter 19.

Finally, on the subject of timber deterioration, there are two points that may in some cases influence the decision on remedial work:

- Plywood and other boards are attractive to the common furniture beetle if the glue used in manufacture is of natural origin. (Most glues in UK manufactured products are now synthetic and not attractive to insects.)
- End grain is porous and will soak up moisture. Built-in joists and studs are therefore particularly vulnerable to rot if moisture is available (Figure 15.4).

MASONRY

In this section, and throughout the book, 'masonry' is taken to include brickwork, blockwork and stonework. More specific terms are used where appropriate.

Materials and workmanship have been more critical in some years than others. Early bricks were hand moulded from local clay, and their variation in strength is understandably high. Early cement mortar was often lean and weak. During peak demand, workmanship declined. The late nineteenth century was a time of declining workmanship, when some buildings collapsed during or very soon after construction. Assuming that distribution curves (Figure 4.4, *page 47*) had the same profile then as now, just as many buildings must have narrowly averted collapse. After this initial brush with danger, in most buildings it is likely that the highest stress concentrations, where the greatest danger lurked, will have relaxed as a result of creep and load redistribution. A modest margin of safety should now have evolved, unless later alterations negated this benefit of time. As a general rule, surviving brickwork should still sustain a compressive stress of 0.42 N/mm^2, provided it is not undergoing progressive movement and

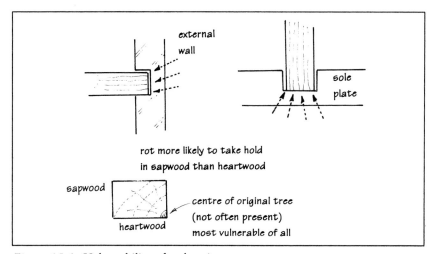

Figure 15.4 Vulnerability of end grain.

has not deteriorated beyond normal wear and tear. (It should not be trusted to take any tensile stress.) However, if there is serious doubt about quality, with a possibility that the proposed work will create at least a local increase in stress, a specialist should be asked to appraise and carry out any necessary testing (subject to conservation rules).

Most of the masonry built after 1930 began life with a more comfortable margin of safety between working stresses and ultimate strength. It is not often necessary to establish its current strength accurately unless the proposed repairs are likely to create a significant (10% or so) increase in stress (see also *Analysis of masonry strength*, Chapter 20).

Although masonry deterioration is inevitable, in most cases it eats into the margin of safety imperceptibly. Unfortunately, for some historic buildings, the imperceptible can accumulate; if it is neglected, there comes a time when deterioration accelerates, and a structure that previously needed a little maintenance suddenly needs rescue.

Deterioration can be said to have one (or more) of three effects on the structural well-being of masonry:

- Straightforward erosion (the wall loses section area; stresses increase).
- Out of plane distortion (forces become eccentric; bending stresses are introduced).
- Increase in slenderness (the strength of the wall is reduced; stresses become more critical).

Wall tie corrosion and mortar deterioration

The steel or iron in wall ties turns to ferrous oxide if it is exposed to oxygen and water. Within the wall cavity, this starts happening on the day of construction (unless the tie is corrosion proof or has a protective covering), because there is a constant supply of oxygen and a regular supply of water through condensation and occasional rainwater penetration. For corrosion to take place in the outer leaf, the protection afforded by the mortar has to be overcome first. Mortar inhibits corrosion by being alkaline, but over time it becomes carbonated in the same way as concrete (Chapter 14), and corrosion begins that day. So failure will first occur either in the cavity or in the outer leaf, depending on which proves to be the more hostile environment. Exposed conditions, especially near the coast, will hasten deterioration in both places. Outer leaf corrosion is hastened if the mortar is lean, weak, contains acidic aggregate (such as black ash), or if the edge of the tie is close to the outside face of the wall, so that the mortar cover is thin. Cavity corrosion is hastened if the outer leaf is porous, fissile or contains empty mortar beds.

Early ties were cast iron, often protected by a coating of tar applied hot. Some of these have survived better than the later ties made of wrought iron and then mild steel, usually galvanized. The standard of galvanizing was unfortunately relaxed in the mid-1960s, and very many of those ties have now failed. (The standard was raised again in 1981.)

Corrosion is almost inevitable in mild steel ties manufactured before 1981. Failure has been known to occur in less than fifteen years. A large proportion of ties fail before thirty years. So unless outer leaf corrosion is already evident from the splitting of mortar beds, the first clue to likelihood of failure is the age of the building.

The main effect of wall tie failure is to increase the slenderness of the wall. Slenderness is defined as effective distance between restraints divided by effective thickness of the wall. Figure 15.5 shows an example of how loss of ties would reduce effective thickness and thereby increase wall

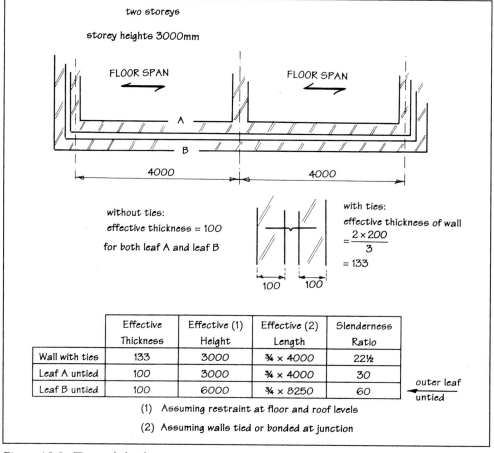

two storeys

storey heights 3000mm

FLOOR SPAN

FLOOR SPAN

A

B

4000

4000

without ties:

effective thickness = 100

for both leaf A and leaf B

with ties:

effective thickness of wall

$$= \frac{2 \times 200}{3}$$

$$= 133$$

100   100

	Effective Thickness	Effective (1) Height	Effective (2) Length	Slenderness Ratio
Wall with ties	133	3000	¾ × 4000	22½
Leaf A untied	100	3000	¾ × 4000	30
Leaf B untied	100	6000	¾ × 8250	60

outer leaf untied

(1) Assuming restraint at floor and roof levels

(2) Assuming walls tied or bonded at junction

Figure 15.5 Ties and slenderness.

slenderness, but that is only part of the story. The outer leaf, by losing its contact with the inner leaf, will also lose the benefit of restraint provided by internal cross walls and floors. The effective distance between restraints would increase considerably in this case, as would slenderness. It would be simple to analyse each wall on the assumption that the two leaves are untied. If the wall remains safe under all loading conditions, wall tie corrosion need not be a problem. (However, a check on forces would be advisable, because the ability to share between leaves will have been lost. Also, rust damage may need to be cut out locally.) In most cases, structural analysis would demonstrate the need for ties.

Figure 15.5 applies to the final condition, assuming total loss of restraint. In practice, tie failure would be piecemeal and the final condition would not be approached until some time after the average life expectancy of the ties had elapsed.

Mortar deterioration

Sometimes, it is the mortar that deteriorates, to the point where it becomes loose enough to be scraped out effortlessly with a blunt tool. This is especially the case with early cavity walling, made when the quality of lime mortar was often poor. Early cement mortars are equally suspect. They were frequently lean and weak. In areas where cavity brickwork was established before World War I (North Essex, for example) and where mortar deterioration has occurred, even if the wall ties are sound, they are no longer tying the leaves together effectively.

Severe mortar deterioration can significantly reduce the compressive strength of cavity or solid walls (typically by 25% or so, but occasionally by 50% or more). It can destroy any tensile or shear strength. One strange outcome of this, fortunately not common, is the dislodgement of bricks by wind suction.

Frost

Some bricks and stones are inherently frost resistant. To others, frost is a hazard. Vulnerable locations include chimneys, parapets, free-standing walls (unprotected by building eaves), brickwork below damp-proof course level and retaining walls, if ground water can percolate through them. On a winter's day, the temperature of a wall's external face may fluctuate around 0°C, alternately freezing and thawing any water it contains. Unless the material is frost resistant, parts may flake off under the bursting pressure of ice, in a modest but persistent imitation of rusting.

There are, in addition, two well-established ways in which the resistance of masonry to frost can be seriously reduced by bad design or workmanship: hard mortar and hard render. Mortar need never, for structural
reasons, be stronger than the masonry units. Usually it can be weaker,
at least to a modest degree, without undermining the wall's margin of
safety. When it is made stronger (nearly always by including too much
cement in the mix), it will often be less permeable than the individual
brick, stone or concrete unit. The effect of this is illustrated in Figure 15.6.

Similar principles apply to render. Rainwater will always soak the render and then (in all but sheltered areas) at least the outer part of the underlying masonry. It then needs to be able to drain and evaporate outwards
efficiently. This can only happen if the wall's permeability increases, layer
by layer, towards the outside face. In other words, the first coat of render should be more permeable than the masonry; the second coat should
be more permeable than the first; the third, if there is one, more than the
second. Finally, paint should not act as a vapour barrier. Sometimes a thin
hard render (perhaps containing a waterproof additive) is applied in the
belief that it will act as a raincoat and keep everything behind it dry. This
never works. Hard render will admit rain through its normal micro-cracks,

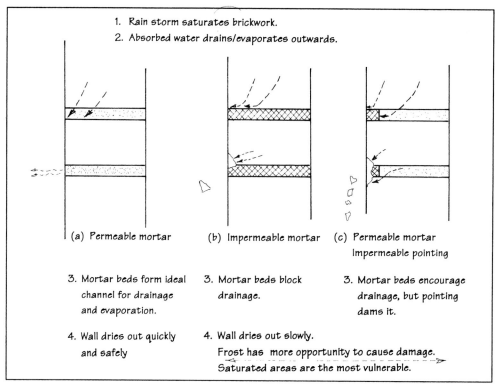

Figure 15.6 Mortar permeability and deterioration.

and eventually through its entire section if the storm is prolonged or severe. Afterwards, the render's low permeability will simply slow down evaporation, keep water in the underlying masonry for longer than necessary, and increase the opportunity for frost to cause damage. Often, the trapped water travels downwards until it collects on a ledge, where damage is intensified. (If the wall is timber, trapped water can instigate rot.) The trapping of moisture and vapour within the wall, both from rain and from internal humid air, is especially harmful to older buildings. Having no barrier or cavity, they rely on outward evaporation as the only safe route for water to take; if this is impeded, there is a very high risk of rot and frost damage from both rain penetration and interstitial condensation.

Glazed bricks and engineering bricks can also be damaged by the vapour they trap behind them.

Crystal growth

Damage similar in appearance to frost action may be caused by crystal growth behind brickwork with a thin face. This creates pressure which leads to the face spalling off. The effect is often most noticeable below cills and copings.

Bowing

One of the secret agents of deterioration is shown in Figure 15.7. Often, continuous strips of timber were used in masonry (mainly brickwork) to improve the bonding or tensile capacity of the wall. These tend to shrink or rot, leaving the wall with a deep horizontal chase, weakening it and creating a permanent bow in it.

Many older buildings contain walls with facing stone or brickwork hiding a rubble core (Figure 15.8, *overleaf*). The rubble is obviously

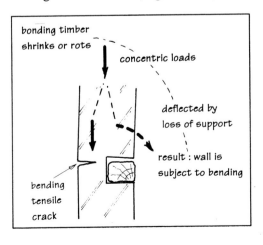

Figure 15.7
Cracking caused by deterioration.

weaker than the facing masonry, but that is of no importance when the wall is first built, because stresses are low. Over a long period, the rubble may consolidate. Minor everyday movements continually break the friction that keeps the rubble stones in place, and any water soaking into the core, either naturally or because rainwater goods are defective, will lubricate the points of contact, further encouraging the tiny but repetitive movements. The result is that the core slowly packs down and presses outwards against the facing masonry. Water in the core may also freeze from time to time, adding to this outward pressure. If the bending moment caused by bowing is high enough to create tension (Chapter 3 and Figure 3.12, *page 39*), then cracking may appear.

Where piers or towers have a rubble core, the bowing caused by consolidation appears preferentially, or entirely, on the outer face, and this increases the width of cracking.

Unexpected weakness: brickwork

Masonry usually performs better in bending than designers give it credit for; but there are occasions when it performs unexpectedly worse, because of weaknesses that are not immediately obvious. These occasions are more common in brickwork than in stonework.

Snap headers

What look like normal headers, bonding the wall across its thickness, turn out to be cut bricks. The lack of contact between inner and outer face

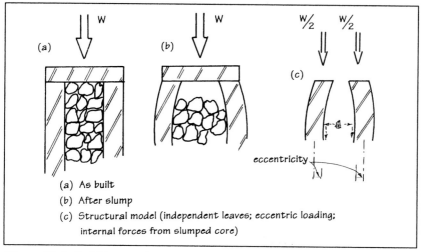

(a) As built
(b) After slump
(c) Structural model (independent leaves; eccentric loading;
    internal forces from slumped core)

Figure 15.8 Rubble core slump.

reduces the wall's capacity to share loads evenly, and it increases slenderness, with the result that stresses are higher than expected and brickwork is weaker. Delamination is a typical symptom.

Poor quality back-up brickwork

Good-quality facings disguise weak and friable commons behind them. Facings and commons are built to a different gauge, so that the courses do not coincide and there is little or no bond between the two. Again, delamination is often the result.

Quality of materials

Quality of materials can be more variable than cursory inspection suggests, especially with nineteenth century brickwork. Underburnt bricks and weak mortar may have been satisfactory for the initial demands made on the building; but, after many decades of alteration and distortion, it may now be working at higher stresses and be in need of help.

Thin piers of brickwork

These may simply be stacks of bricks with no bonding. Bay windows, for example, sometimes contain load-bearing mullions with surprisingly high slenderness. Nevertheless, they usually serve their purpose for many years until challenged by decay or an introduced structural problem.

Chimneys

Chimneys usually consist of a perimeter shell of brickwork restrained by the dividing walls between flues. In practice, the dividing walls are often unbonded and their mortar is degraded by the acid flue gases, so that they provide little or no restraint to the thin shell.

Thick walls

Thick walls may be hollow or may contain flues and air vents.

Stone

Stonework is prone to most of the agents of deterioration that assail brickwork, and a few more. As a natural material, its properties are variable, especially its resistance to decay. This does not cause frequent structural problems, because stonework is usually strong and massive compared with other traditional building materials. Therefore it has a good margin of safety, protecting it from the increased stresses and distortions that arise from deterioration, although it is certainly not indestructible.

Sedimentary stones should nearly always be laid on their bedding planes. In other words, the horizontal planes in which they were deposited, regardless of subsequent geological movement, should be the planes in which they are horizontally bedded into the building. If, instead, the bedding planes are set parallel to the building facade, for example, the stone may delaminate or become pock marked.

All stones should be carefully cut at the quarry and carefully dressed for their particular role. Inept working will hasten deterioration, usually by presenting damaged (bruised) faces to the elements.

Most stones contain joints and fissures, which are vertical or inclined discontinuities formed by shrinkage or early distortion of the original rock, in many cases subsequently filled by solution. These may or may not be areas of weakness. Weak joints (vents) may allow differential erosion.

Salts

Erosion can be caused by solution, as rainwater containing weak acids washes across the face of the stone. Atmospheric carbon dioxide, dissolved into the rainwater, increases the rate of solution of any stone containing calcium carbonate. Calcium carbonate is the main constituent of limestone and is the cementing material of calcareous sandstone. (The solid calcium carbonate reacts with the dissolved carbon dioxide to form calcium bicarbonate, which is soluble.)

Historically, the effect of polluting gases has been far more severe than the effect of carbon dioxide. Acid sulfur, once common in coal-burning industrial areas, also erodes calcareous stone. Reactions between dissolved sulfur and calcareous material create a hard impermeable skin of salt (sulfate and sulfite) that traps water behind it, like a hard render, until evaporation or frost action causes blistering and exfoliation. The salt skin may become more established in sheltered parts of the building than in parts regularly washed by rain.

Other soluble salts cause damage by crystallization. The salts may come from the building itself (the stone or mortar or backing material), from products of decomposition washed onto the stonework, or from the atmosphere. If the salts crystallize on the surface, they cause efflorescence, which is harmless if it forms a loose powder, but not if it forms a more tenacious skin. If the salts crystallize below the surface (cryptoflorescence) they partially fill pores. This may lead to damage by evaporation or frost. Salts introduced to the outside face of a stone sometimes migrate inwards under the pumping action of a drier warmer indoor atmosphere, so that any cryptoflorescent damage appears on the internal face of the wall.

Sandstone

Sandstone can be damaged if limestone is used on the same building at a higher level. If rainwater can run or splash from the limestone onto the sandstone, the run-off water will contain a solution of calcium sulfate that attacks sandstone. If the limestone is magnesian, the water may also contain magnesium sulfate, which is all the more deleterious to sandstone, sometimes causing severe decay. In fact, magnesian limestone will cause other limestones to decay in the same circumstances.

Some parts of the UK, notably the Bristol area, are close to both sandstone and limestone quarries. The two materials have been mixed in a number of buildings, often for good reasons, but sometimes with disappointing long-term results.

Cutting or grinding sandstone is hazardous, because it releases silica dust. Breathing in silica dust is extremely harmful and can lead to silicosis. It is for this reason that silica grit blasting of masonry surfaces was banned some time ago. Any repairs involving sandstone have to avoid this severe hazard.

Weathering

Resistance to weathering depends on many factors beyond the scope of this book. Experts can advise on the subject and can also judge or, if necessary, test for the durability of any type of stone that is being considered for repair work.

Earth

Earth, simply mixed and compacted but not fired, was in continuous use from the early eighteenth until the early twentieth century for the walls of small buildings, such as houses and barns, in sheltered areas. Earth buildings may not always be recognized now for what they are, if they are covered by plaster and render. In some cases, a brick face may have been added.

The basic material of earth walling was either clay or chalk. It was mixed with other soil – silt, sand, gravel – and fibre, such as chopped straw, and was formed into walls *in situ* by ramming the mixture between shutters. Building usually progressed in fairly short lifts of 600mm or so at a time, with a short pause after each lift for consolidation, strength gain and shrinkage, before continuing upwards. There were many regional variations. In East Anglia, most earth walling is clay lump. The clay was mixed and formed into blocks (lumps) in moulds and, after curing, built up in units like blockwork, using mud as mortar. In Lincolnshire, *in situ* clay was built around a lightweight timber frame, a method referred to as 'stud and mud'.

It is impossible for anyone responsible for repairing or maintaining earth walling to escape from its two severe drawbacks: low strength and a durability that is adequate if carefully maintained but fragile if neglected, which sadly it often is.

Materials for earth walling were mixed and used by skilled craftsmen, and there is little doubt about the quality of workmanship in any surviving buildings. However, earth walling is difficult to classify by strength, because each site represents a unique blend of local materials. It is safe to say that no other building material is expected to work so close to its failure stress. That means deterioration has to be watched anxiously. Water and frost are the main enemies. Saturation very seriously weakens the material. Clay can approach its liquid limit; chalk can become putty. This need affect only a small band within a wall to cause failure. A healthy and well-maintained clay lump house collapsed after standing for two weeks in shallow flood water. Less catastrophic swings in moisture content can cause fractures in clay walls, since they shrink and swell in the same way as their parent material (although usually to an extent that has been modified by the mixing). Chalk does not shrink or swell, but it is susceptible to frost. Swings of moisture content in chalk may encourage disintegration through frost damage.

These problems were always acknowledged by the craftsmen, who made the walls thick, their panel sizes modest and openings in them small, often with heavy, strength-reimbursing frames. This avoided high stresses. They protected the walls with wide eaves, a substantial masonry base (not necessarily damp proof, but with its top edge high enough above ground level to keep the earth wall it supported beyond the range of rising damp) and a soft renewable render. This kept deterioration at bay. The inherent benefits of these measures can be defeated by:

- unwise alteration
- poor maintenance
- a change in the environment.

Alterations that would be modest in masonry buildings can cause unacceptable damage to earth walls if they increase loading (by inserting openings, for instance) or increase slenderness (by removing cross walls, for instance).

Earth walling is easily damaged by impact. It can also be weakened by vermin that burrow into it, birds that nest in it, and even cows that lick it. Occasional patching should be regarded as being as normal as painting (if perhaps not quite as frequent). Neglect of such maintenance can be swiftly punished. Renewal of render should also be accepted as a normal maintenance item. Inevitably, modern cement render has been applied to many earth walls, with sad consequences. Hard render does even more harm, even more quickly, to earth than it does to masonry.

The sensitivity of earth walling makes it vulnerable to deterioration or structural failure, even when it has been well maintained, if the environment changes for the worse. The danger of flooding has already been mentioned. Vegetation should not be allowed near earth wall buildings, if it is likely either to exaggerate seasonal movement of the soil or to take root within the fabric of the building. Busier roads mean increased traffic vibration. This is seldom a problem with other materials, but with earth walls, the margin of safety can be eaten up by minor troubles.

Before repair or maintenance can be specified, the designer has to make an assessment of the structural condition of the building. This should take into account any previous alterations, lack of maintenance and changes to the environment. Above all, nothing should be done to increase the risk of saturation or to interrupt the free movement of vapour through walls. When it comes to altering the structure itself, nothing should be done to increase stress or slenderness. With almost any other building material, it is normal to make modest alterations, such as inserting small openings and providing lintels or beams to carry the loads above them, without formal structural reappraisal of walls or foundations. Not so with earth walling, which should never be altered, except to improve its strength or resistance.

PART THREE

Options

Chapter Sixteen

Purpose of repair

It might be thought that the purpose of repair is to restore the building to its original structural condition, but that is not always necessary. In some cases, lesser attention may be sufficient to make it safe and serviceable for its future purpose. On other occasions, full restoration might be desirable, but shortage of funds or a delay in securing funds restricts current attention to temporary (running) repairs with a more limited purpose. This chapter suggests a method of focusing on the purpose of repair to clarify the options, and it emphasizes the need to communicate purpose to owners and funders.

Type of cause

It can help to focus on the purpose of repair by classifying the cause of damage under the following headings:

- single incident
- cyclic movement
- deterioration
- progressive movement.

Single incident

A single incident is one whose cause no longer operates. This greatly simplifies the purpose of repair. If there are no financial or other constraints, it is simply a question of correcting any loss of internal or external equilibrium (Chapter 2), and tidying up cosmetically.

Any long-extinct cause of damage can now be regarded as a single incident and dealt with accordingly. Foundation settlement used to be gross by today's standards but, provided building and soil are now in equilibrium, there is no point in replacing or improving the 'failed' foundations. (If the building is about to be altered, and fresh loading will be imposed on it, that would be a different matter; such cases warrant careful analysis.)

Cyclic movement

Thermal expansion and contraction of walls, and seasonal movement of foundations, are examples of cyclic movement. Sometimes (Chapters 9 and 12) there is a slight progressive element to the damage unless it is repaired regularly. Nevertheless, the limits of movement can usually be forecast for the medium term. Repairs to cyclic movement will have one or more of the following purposes:

- Improve appearances for the time being, by carrying out running repairs (and monitor future movement and damage).
- Restore any restraint destroyed by the movement (such as wall-to-wall connections; Figure 12.3c, *page 119*).
- Provide restraint to cope with future movement.
- Provide joints to cope with future movement.
- Very occasionally, severe distortions may need to be redeemed.

Deterioration

Deterioration can, by secret accrual, lead to sudden damage. More often, it diminishes strength or robustness, which heightens the consequences of other active causes. It may be necessary, as part of any repair specification, to compensate for past loss of strength, or to slow the rate of deterioration so that it will not significantly contribute to fresh problems in the foreseeable future.

Progressive movement

Progressive movement remains progressive unless and until the underlying cause is eliminated. The basic options for dealing with progressive movement are:

- Do nothing yet (and allow the problem to run its course, then repair as necessary).
- Mitigate further damage (and reduce the severity of the problem as it runs it course).
- Prevent further damage (by arresting the problem).

Chapter 9 describes many causes that would, in the right circumstances, be suitable for the middle option. These circumstances are right when:

- it is safe to let the cause run its course
- it would be expensive to arrest movement altogether
- it is relatively simple to reduce the level of damage while the cause still operates.

The concept of mitigation is simple, but the practice is difficult, because prediction (which governs the first and third qualifiers) is usually uncertain. Practice should therefore be attended by monitoring and contingency plans.

There was a decade, preceding insurance cover, when an epidemic of heave badly damaged many houses in southeast England. In several cases, predictions for total movement were gloomy, but funds for underpinning were unavailable. Mitigation was a forced option. Strapping beams (Figure 17.4, *page 160*) and local reinforcement (Figures 20.6, 20.15 and 20.16, *pages 207, 218 and 220, respectively*) were often used, and it can now be said that they achieved a high success rate. Most of the houses repaired in this fashion – in fact the comfortable majority – survived the remaining heave movement with negligible or cosmetic damage. This experience does not alter the fact that every fresh incident, when first discovered, faces uncertainty, inconvenience and temporary loss of the building's value if mitigation is the preferred option.

Figure 16.1 (*overleaf*) is a flow chart that breaks down the three-option choice into a series of decisions. The chart is an oversimplification, in particular ignoring non-technical restraints such as conservation. Its limited purpose is to emphasize the range of choice and the need to be clear about the consequences of each action, or inaction.

Constraints and opportunities

Part of the skill of the repairer lies in early recognition and definition of constraints (such as the need to keep the building in use) and opportunities (such as solving non-structural problems during the structural repair contract, and sharing overheads). These should form part of planning. One point is sometimes more obvious to owners and funders than to their professional advisors: repair competes for funding with other financial options, including the option of leaving the money in the bank.

Communicating the purpose

Owners need to be informed about the technical constraints on their own priorities. This is especially so when technical proposals contain a high degree of uncertainty, as is often the case, for example, with mitigation.

If they are to keep the owners adequately informed, repairers must retain a clear understanding of the purpose of the repair – what it aims to achieve in technical terms. This must also be accurately expressed and communicated; otherwise the parties to the repair contract will all conceive their own versions and, as time goes by and events unfold, each individual version may evolve in isolation until the contentious reckoning. A technically good job is easily spoiled by poor communication.

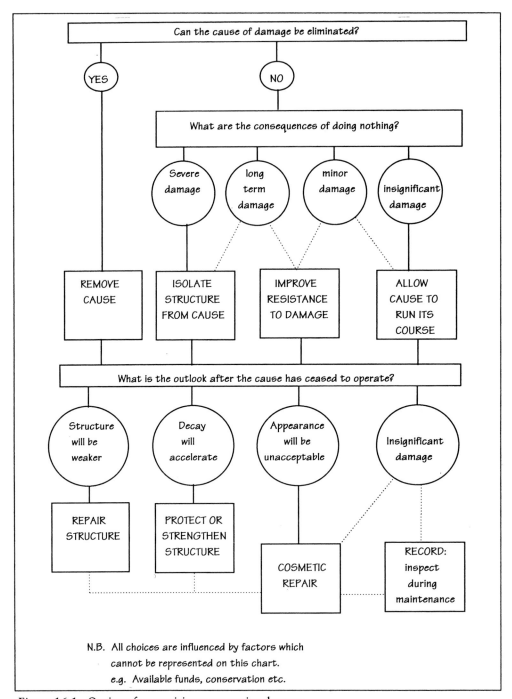

Figure 16.1   Options for repairing progressive damage.

Chapter Seventeen

Foundations

Foundation damage is more difficult to solve and more costly to remedy than most other defects. The problem usually lies in the soil, which is found to be inflicting its own volume changes on the building it is supposed to be supporting. This chapter introduces two criteria that also influence options – health and safety and unforeseen ground conditions – and then considers the broad options for repair.

Health and safety

Figure 17.1(*overleaf*) shows examples of hazards encountered during underpinning. The stability of soil supporting a foundation load always depends to some extent on the restraint provided by the soil above and around it (Figure 3.10, *page 38*). Removing this restraint will, therefore, reduce the margin of safety (Figure 17.1a). The consequences can be disastrous. In the case of clay soils – stiff clay in particular – these consequences are not immediately apparent, because, as hand excavation proceeds, the soil appears to be strong and unyielding, and there is no outward sign of the inevitable weakening that begins as moisture drains towards the exposed face. You need more than common sense to anticipate failure in this case. The enthusiasm of site operatives to make the best possible progress can tempt them to sail too close to the wind, without understanding the risks they are taking.

When footings are undercut (Figure 17.1b), individual units such as bricks and stones may come loose. They must, of course, be stabilized before anyone enters the space below. This can be a tedious process. If the units merely loosen without falling, and are not tightened up, the underpinned building may experience a larger than expected bedding down.

If the building is fragile, the designer may need to warn the contractor that the potential for damage and injury is greater than usual. In addition to providing the usual struts and sheeting within the excavation, it may be necessary to improve the building's robustness before work starts.

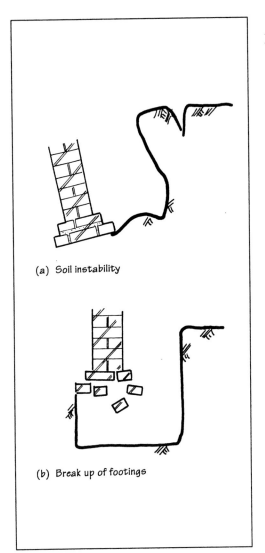

Figure 17.1
Underpinning hazards.

(a) Soil instability

(b) Break up of footings

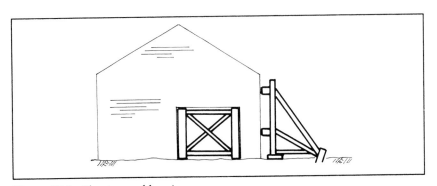

Figure 17.2   Shoring and bracing.

Precautionary measures (Figure 17.2) may include:

- shoring it against lateral movement
- inserting props or needles to prevent vertical movement
- bracing large openings
- sometimes, in addition, strapping corners or temporarily repairing cracks to prevent any loss of equilibrium.

Unforeseen ground conditions

It is impossible to have perfect foreknowledge of ground conditions, because this would require an investigation so comprehensive that its cost would overshadow the remedial works contract. Every contract should allow for some variation to the work, in the light of unexpected conditions. The question is, how much of the unexpected is it reasonable to expect?

Figure 17.3 illustrates the well-known association between cost of investigation and cost of variations. The cost that matters is total cost. That comprises: investigation cost, plus tender cost (or in the absence of a tender procedure, the quote, estimate or budget cost), plus cost of variations. There is a level of investigation that leads to minimum total cost. This ideal level is theoretical and unforeseeable, but, given a free hand, an experienced designer should come close to it on most occasions. Unfortunately, commercial or political pressures often drive down the actual investigation cost and conspire to disguise the influence of this decision

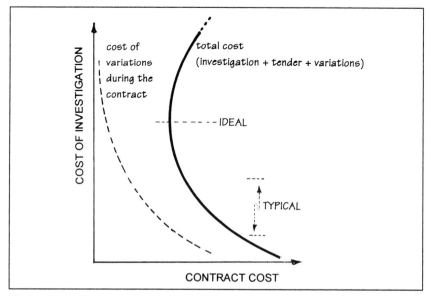

Figure 17.3   Investigation and contract costs.

on the increase in overall cost. Restricting the investigation can also limit the designer's opportunity to reduce risks to health and safety. This is particularly true if ground conditions are variable.

Investigation obviously should be controlled, regardless of whether or not its budget seems adequate, so that the best value for money is obtained. Best practice includes:

- getting the most out of the low-cost tasks, such as consulting records (Figure 7.2, *page 66*)
- weighting resources to eliminate the potentially more expensive areas of ignorance
- organizing the remedial contract itself to supply early additional information.

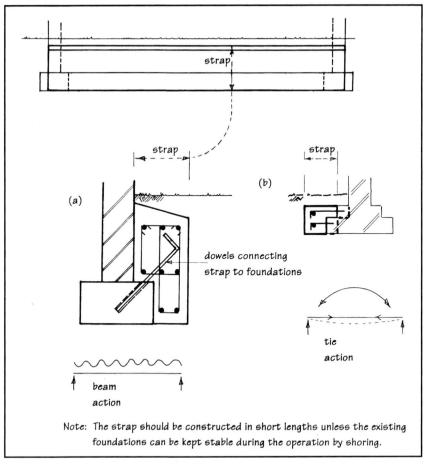

Figure 17.4 Foundation strapping.

Options for repair

Strapping beams and ties

Figure 17.4a shows a typical reinforced concrete strapping beam, which converts an unreinforced footing into a powerful beam. It can be used to bridge across loosened soil. It can also be used to strengthen and stiffen the building during limited recovery or heave. In the latter role, the beam can only mitigate further damage; it cannot be designed to prevent it altogether, because the future pressure from the swelling clay is unknown.

In buildings with shallow footings, there may not be enough vertical space to fit a usefully stiff beam, except by underpinning. A shallow strap (Figure 17.4b) may still be of benefit as a tie, encouraging the wall to arch over weak soil. For this to work, the wall should not be heavily pierced by openings; there should not be a *low bond* damp proof course that might allow slippage in response to the horizontal component of the arching force; and there should be enough solid brickwork below any cavity for a key into the tie.

Pre-stressing may be used as an alternative to simple reinforcing, provided the substructure can withstand the forces applied during the act of prestressing.

Strapping is an option that requires engineering judgement, because its design is empirical.

Types of underpinning

Figures 17.5 to 17.8 show several methods of underpinning. The choice between them should be based on safety, economy and practicality.

Continuous strip (Figure 17.5, *overleaf*) is simple, and economic if depth does not exceed 2 metres; but it is increasingly uneconomic as depth increases below 2.5 metres. It may serve a dual purpose, as new foundations and as a barrier to roots penetrating beneath the building (but see Chapter 25 for warnings about root barriers).

Pad and beam (Figure 17.6, *page 163*) is simple and economic for underpinning depths between 2.5 and 4.5 metres. If the ground contains obstructions (congested services, large boulders or, in made ground, lumps of concrete etc.) then this method may be the only practicable form of underpinning.

Pile and beam (Figure 17.7, *page 164*) is likely to be more economic than both continuous strip and pad and beam if underpinning depth exceeds 4.5 metres, and there are no underground obstructions. There are several options for connecting piles to beams, giving scope for economy

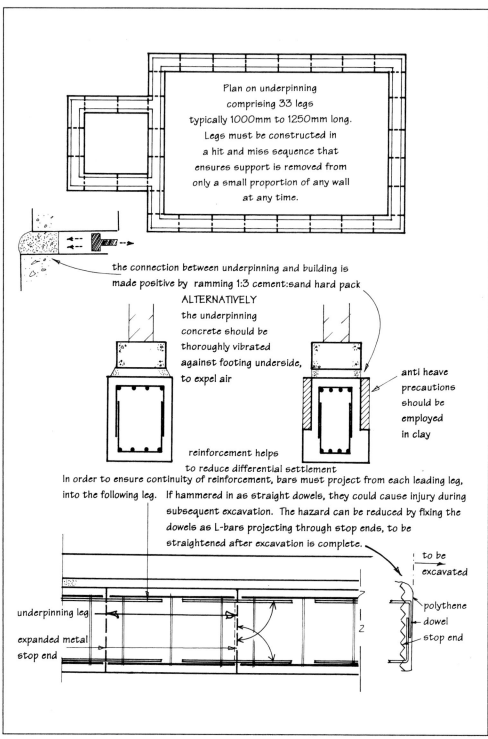

Plan on underpinning
comprising 33 legs
typically 1000mm to 1250mm long.
Legs must be constructed in
a hit and miss sequence that
ensures support is removed from
only a small proportion of any wall
at any time.

the connection between underpinning and building is
made positive by ramming 1:3 cement:sand hard pack

ALTERNATIVELY
the underpinning
concrete should be
thoroughly vibrated
against footing underside,
to expel air

anti heave
precautions
should be
employed
in clay

reinforcement helps
to reduce differential settlement

In order to ensure continuity of reinforcement, bars must project from each leading leg,
into the following leg. If hammered in as straight dowels, they could cause injury during
subsequent excavation. The hazard can be reduced by fixing the
dowels as L-bars projecting through stop ends, to be
straightened after excavation is complete.

to be
excavated

underpinning leg

expanded metal
stop end

polythene
dowel
stop end

Figure 17.5 Continuous strip underpinning.

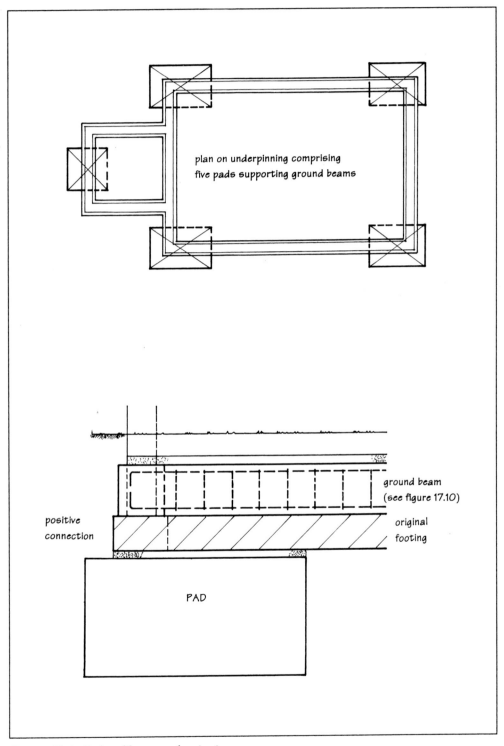

plan on underpinning comprising
five pads supporting ground beams

ground beam
(see figure 17.10)

positive
connection

original
footing

PAD

Figure 17.6   Pad and beam underpinning.

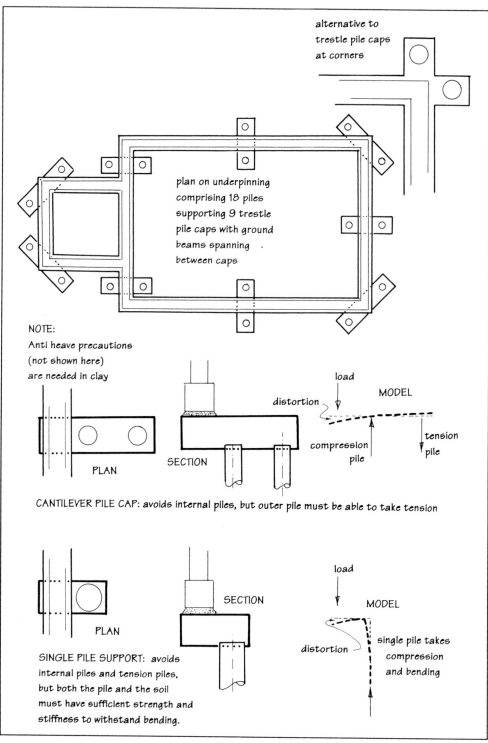

Figure 17.7   Pile and beam underpinning.

according to building layout and site conditions. The main plan shows the standard pile cap or trestle straddling twin piles. This is not always the most economic, or convenient, arrangement. Three other options are illustrated.

Piled raft (Figure 17.8) is a version of piled underpinning, usually the most economic if internal walls and ground floor need to be underpinned, because the raft element forms the base of the new floor. The design can usually be made flexible, with license to amend pile positions to suit the building's internal layout and small underground obstructions.

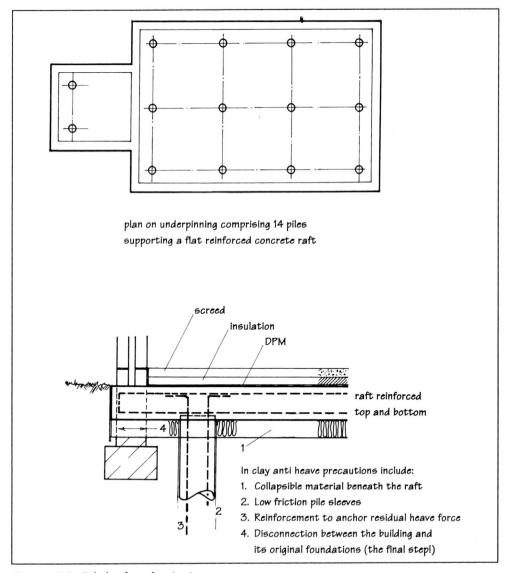

plan on underpinning comprising 14 piles
supporting a flat reinforced concrete raft

screed
insulation
DPM

raft reinforced
top and bottom

In clay anti heave precautions include:
1. Collapsible material beneath the raft
2. Low friction pile sleeves
3. Reinforcement to anchor residual heave force
4. Disconnection between the building and
   its original foundations (the final step!)

Figure 17.8   Piled raft underpinning.

Types of pile

This section discusses technical issues relevant to choosing the right pile for the ground conditions. It strays a little beyond the discussion of soil behaviour in Chapter 3, and may be omitted by readers needing only an elementary appreciation of underpinning.

In clay, most underpinning piling is augered, as the method is vibration free and produces a pile that relies on both its shaft and its base to support loads. This is economic in the use of materials; in fact, shaft friction usually provides most of the working capacity of an augered pile. If water-bearing seams are encountered during the construction, they should be sealed by sleeving. This is relatively simple near ground level, but difficult and expensive below four metres. Continuous flight auger (CFA) piles may be cheaper to install in clay with deep water-bearing seams. They are also worth considering for use in soft clay, as the method does not leave the bore unsupported; during installation, the auger remains in place, preventing the sides of the shaft from collapsing or necking. The auger is withdrawn as the concrete is delivered down its stem, and stability thereafter relies on the mass of the concrete. The shaft of a CFA pile cannot be visually inspected, as it can for open augered piles, and integrity testing should be carried out on the finished product to ensure that no faults have been built in.

Driven cased pile construction is not prone to problems with ground water, but the driving can cause shock waves that are sometimes severe in dense or medium-dense soils. Predrilling a pilot hole to a moderate depth, and then driving from the bottom of the pilot hole, can reduce the effect of these shock waves, and will usually allow a driven pile solution to be adopted even in damaged or sensitive buildings. Driven piles are advanced until a predetermined set (mm penetration per blow) is recorded. Knowing the energy required to achieve the set allows the design load to be predicted by formula. Given the inevitable variations in ground conditions, this means that the final length of piling will vary from the design estimate, because the soil varies – always! The design should be reviewed if early driving suggests that such variation will turn out to be large. In some soils, notably silt and chalk, the action of driving temporarily breaks down soil resistance, so that installation needs much less energy than the formula assumed; it follows that the mm per blow record will then seriously underestimate the true static bearing capacity. If operatives attempt to achieve the standard set, in these conditions, they are likely to produce uneconomically (and unnecessarily) long piles. In silt and chalk, an alternative is to drive each pile to a designed length, based on soil test data, and if necessary to confirm its capacity by test loading.

(The dynamic action of driving raises pore water pressure in silt and chalk, which weakens the soil locally and temporarily, so that the piles fail to set until they have reached abnormal depths or a different stratum. A reasonable period should be allowed to elapse, before testing, to allow the temporarily high pore water pressure to dissipate. Redriving to a set can sometimes be a realistic alternative to testing, provided the delay for dissipation is observed, and provided the redrive achieves an immediate set, before it raises pore water pressure again and the pile 'runs away'.)

In contaminated ground, piling may be chosen in preference to the more hazardous hand digging. Augered and CFA piles would bring contaminants to the surface. If the contaminants are particularly hazardous, driven piles might be considered safer, provided the contaminants are not likely to be pushed towards an aquifer by the pile driving. Advice from geotechnical or environmental engineers should be sought.

Mini-piles are small-diameter (typically 100mm to 150mm) driven piles. They are discussed in the following chapter. They are quite acceptable for underpinning schemes of the type illustrated in Figures 17.7 and 17.8, provided they can achieve the required capacity. (If the piles are long compared with their diameter, slenderness can seriously reduce capacity.) If the existing footings are in good condition, driven mini-piles can sometimes be drilled through them (at a small angle to the vertical) and then be connected directly into the building's substructure, thus saving the cost of pile caps.

Protecting the formation

Water must not be allowed to soften any of the soil that will be supporting the building when the underpinning is complete. In the case of open augered piles, a flooded shaft would usually have to be abandoned altogether and the scheme redesigned. Figure 17.9 (*overleaf*) illustrates dangers and solutions associated with hand-dug underpinning.

Forming beams

Figure 17.10 (*page 169*) shows a safe method of forming the ground beam in preparation for a pad and beam or pile and beam scheme. In this example, the beam is placed above original footing level. If the footing were shallow, the beam would probably have to go underneath (using temporary footing pads meanwhile to support the props). The beam's position, in each case, should be individually designed for safe construction and maximum economy.

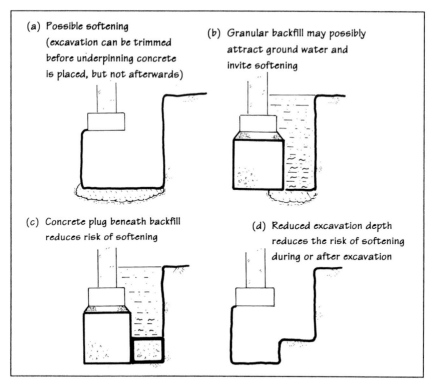

Figure 17.9 Protecting the formation.

Partial underpinning

Partial underpinning will be successful if all areas of the foundations previously moving are rigidly supported by the underpinning and all areas
that are not underpinned can be relied upon to remain stable. If that were
a simple judgement to make, there would be no difficulty in agreeing the
limits of partial underpinning. There is often sufficient uncertainty, however, to provide grounds for disagreement.

If level monitoring has been carried out, the area of instability may be
well defined. However, it is rarely worth monitoring for that purpose
alone. Without monitoring, judgement must rely on other clues, such as
distortions, soil conditions and damage pattern. Observation of damage pattern alone would usually lead to overestimation of the extent of
instability, because buildings redistribute forces to accommodate foundation movement, and stresses are increased within areas above stable
ground. (This is discussed in Chapter 4 under *Soil–structure interaction*.)
There are two other considerations:

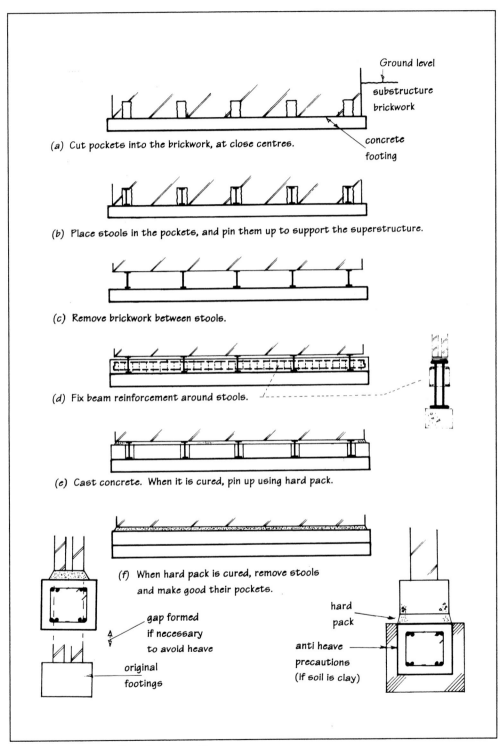

Figure 17.10 Installing the ground beam.

- The non-underpinned area may be prone to seasonal movement if foundations are shallow and the soil is plastic.
- The non-underpinned area may be prone to creep settlement, if the building is less than ten years old or is founded on made ground, peat or soil that is loose or soft.

Rarely is it value for money to carry out a total underpinning as a precaution against seasonal movement or creep. If the underpinning design is bad enough, it is possible to exaggerate any damage that might arise from these causes. Normally, this is avoided by designing a transition between underpinned and non-underpinned parts. There is a three-stage method for doing this:

1. Establish the area of soil instability.
2. Decide on the most appropriate (economic/safe) method of underpinning the unstable area. Usually the choice is between continuous strip, pad and beam, or pile and beam. A piled raft is not usually adaptable to a partial underpinning scheme.
3. Design a transition to extend from the boundary of the unstable soil into the stable soil, for a reasonable distance.

In the case of a continuous strip underpinning, the transition usually consists of a succession of legs of diminishing depth, the final one being perhaps no deeper than necessary for practical purposes (Figure 17.11a). In

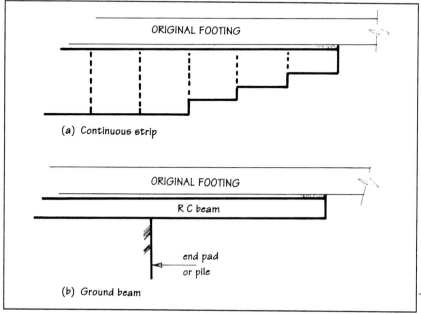

(a) Continuous strip

(b) Ground beam

Figure 17.11   Typical transitions.

the case of pad and beam or pile and beam underpinning (b), the reinforced beam is usually continued for a short distance beyond the edge support.

It is good practice to end the transition below a strong point in the building. If, for example, a succession of standard regular steps (Figure 17.11a) would terminate the transition beam at the leading edge of a large opening, leaving the opposite side on its original footings, it would be wise to lengthen the transition so that both sides are supported (Figure 17.12). The reasoning is that openings represent weaknesses that attract cracking as a result of minor differential movement. A transition supporting only one side might exacerbate this weakness; a transition supporting both sides will reduce it.

It is worth noting that if the cause of damage (subsidence or heave) were to spread to parts not underpinned, the transition will not help the building for very long. The remedial proposals may need to include precautions, such as vegetation control or drain improvements, to reduce any such risk (Chapter 25).

Jacking

Some clients assume that underpinning will correct distortions, but it merely freezes them. The original shape of a building can be restored (to an extent, depending on circumstances), by jacking the building off the completed underpinning, in accordance with the following sequence:

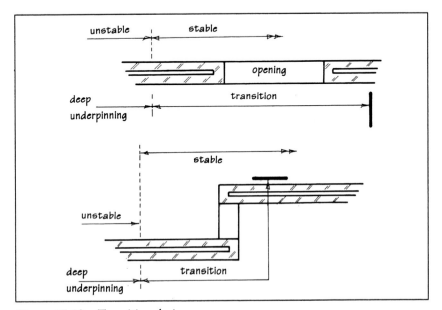

Figure 17.12   Transition design.

1. Measure existing distortions.
2. Decide on the intended profile (usually determined by the extent to which it seems possible to iron out existing distortions).
3. Prepare the building to accept the temporary forces created by jacking.
4. Carry out the jacking, continuously reviewing progress and amending the intended profile if necessary.

Jacking attempts to reverse, almost instantly, distortions that may have taken months or even years to form. Building materials are more stiff when the loading is short term; furthermore the positions of jacking points will not be a mirror image of vanished support. These practical difficulties inhibit total success. In the case of a small extension that has simply tilted away from the rest of the building, re-levelling by jacking may achieve nearly 100% reversal. However, a building with complex distortions, especially if from more than one incident, will have to accept a less satisfying compromise.

Correction of distortions can often be encouraged by cleaning out or enlarging existing cracks and by removing frames that might resist the jacking forces – or, of course, be damaged by them. Since jacking exerts vertical force in an effort to correct rotational distortion, it is also necessary to consider whether horizontal forces need to be applied or resisted during the operation. For these reasons, jacking is a specialist task and needs to be planned and supervised by an experienced engineer who is able to make adjustments and compromises without reporting back.

To make an obvious point: jacking need not rely on underpinning to provide a firm base. If the building is stable, and does not require underpinning, but the owner does wish to reduce its distortions, temporary bases can usually be designed to support the jacks.

As well as correcting distortions, jacking can be used in the following circumstances:

- *Pre-loading:* confirming the bearing capacity of piles or pads and accelerating bedding down. (A specialist should confirm the suitability of any proposal; it would not work in all soils.)
- *Separation:* creating a gap, after the underpinning is in place, between the original foundations and the soil. (Where clay is heaving, or expected to recover, this can be used as a substitute for anti-heave precautions, often achieving worthwhile savings in cost.)
- *Leaning walls:* reducing severe leans by rotating the wall into a safer profile. Usually the rotation is effected by gripping the wall within a stiff cage of temporary shoring and jacking against the cage. Methods using permanent shoring (buttressing) are sometimes equally suitable. (The wall's natural resistance to lateral movement has to be temporarily

overcome, before jacking starts, by removing its connections with floors, roofs and cross walls. It is also necessary to form a horizontal hinge for the rotation to develop at the intended level. After jacking is complete, connections are restored and the hinge is made good.)

- *Relocation:* where the building is to be moved to another site, and where relocation is by transport rather than skidding, jacking would be used to lift the building off its existing foundations (stiffened if necessary) and down onto the transport. At the new location, jacking would be used to lift the building off the transport and down onto the prepared new foundations. (These must be stiff, otherwise the building will undergo a second 'initial' settlement.)

Chapter Eighteen

Ground floors

Severe structural damage to ground floors is not common. On the other hand, the cost of correcting even modest defects is high, mainly because of the inconvenience and difficulty of preparatory work. If the level of damage does not warrant the cost of a perfect cure, running repairs, entailing less preparation, may well be the preferred option. This chapter discusses a range of remedies.

Ground-bearing concrete floors

If loose hardcore has permitted abnormal settlement, there are many possible remedies to choose from:

- Do nothing; accept the loss of performance.
- Disguise the movement by renewing the screed or applying a new self-levelling compound above it.
- Replace the hardcore and everything above it.
- Replace the ground-bearing slab with a new suspended slab.
- Stabilize the hardcore by grouting (in the right circumstances, pressure grouting may be used to raise a floor slab that has settled or subsided).
- Support the slab on mini-piles.
- Use a combination of grouting and mini-piling.
- Underpin the ground slab and foundations in combination (e.g. raft).

Unless the hardcore contains collapsible or degradable material, its settlement should be substantially complete within ten years. Hence, repairs are often dealing not with active movement but with its legacy. In most cases, renewal of the screed, or application of a self-levelling compound on top of the existing screed, should be satisfactory.

If the hardcore is heaving as a result of sulfates within it or, less commonly, sulfates conducted into it from the subsoil, then it is not easy to be optimistic about its future performance. Expansion will continue until all available sulfates have been converted, and neither time nor movement is predictable. Unless movement can reliably be said to have stopped, it will be necessary to break out the floor and replace the offending hardcore with inert material, before laying a new floor. In some cases, it may be more economic to replace the existing ground-bearing floor with a suspended floor, leaving sufficient space beneath the soffit of the new one to allow for future expansion of the hardcore.

Likewise, a suspended floor may be the cheapest cure for hardcore that is still settling. The cheapest way of supporting the new floor slab would be to provide a bearing for it off the existing foundations. This would obviously add to the bearing pressure. Figure 18.1a shows the new suspended floor supported on 'hit and miss' ledges cut into the existing brickwork. If the existing brickwork is fragile, it may be easier to provide a new bearing (Figure 18.1b). In both cases, the load on the soil would be increased and, in fact, the increment would be eccentric; hence the increase in bearing pressure would be disproportionately higher. If there is not a comfortable margin between existing foundation bearing pressure and soil bearing capacity, then the new floor could be supported on a downstand (Figure 18.1c) made wide enough to minimize its own settlement. Its width and position, where there is scope, should also take into

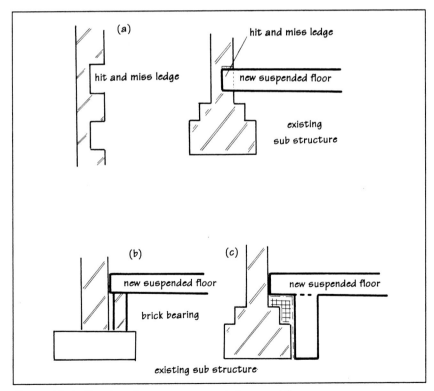

Figure 18.1 Bearings for a new suspended ground floor.

account the potential imposed settlement on the adjacent foundations. It is a more expensive detail than the others, not least because the excavation for the downstand would be close to existing foundations. For safety reasons, it should be constructed only in short lengths.

Stabilizing by grout (Figure 18.2) is a specialist repair. It works only if the material being grouted has a much coarser particle size distribution (Figure 3.1, *page 23*) than the grout itself, and if it is reasonably and consistently permeable; otherwise flow is obstructed, with the risk that a proportion of gaps and voids will be left unfilled. A high ground water level and chemically contaminated hardcore also preclude grouting.

The purpose of grouting is usually restricted to stabilizing the loose hardcore or subsoil, and the material most often used is a mixture of cement and pulverized fuel ash (PFA). The latter prevents the solids from settling before hydration is complete and thereby encourages consistent quality. Additional fillers or additives may be used, appropriate to the conditions. If it is also necessary to strengthen the floor to take heavier loading, or to re-level it by injecting the grout under pressure, other materials or methods may need to be considered. But for the purpose of stabilization only, the steps in designing a grout scheme are as follows:

- Confirm that grouting is feasible.
- Choose a suitable mix.
- Choose injection points and rates of injection to ensure that the grout adequately fills all gaps and voids without spreading beyond the area to be treated.
- Protect service ducts and drains that might be accidentally flooded by grout – and ensure that such vulnerable areas are checked during the work.
- Apply the grout, adjusting the mix if needed as its performance registers.
- Continually check grout spread.

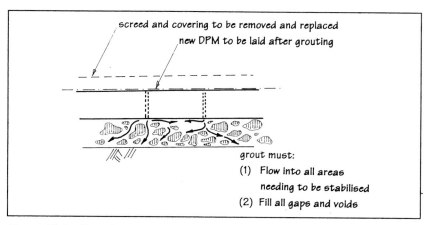

Figure 18.2   Grouting.

- Clean off splashes.
- Re-establish the damp-proof membrane, ensuring that all gaps in the barrier against water are closed.

Mini-piling does not stabilize the hardcore. It transfers the weight of the ground slab and all that it supports to a lower, stable stratum (Figure 18.3).

Mini-piling may be used on its own or in combination with grouting. One advantage of a combined scheme is that it would not be essential for the grout to fill 100% of all voids in loose material. However, it should, without difficulty, fill the upper voids, including any gaps that may have opened between slab soffit and hardcore (Figure 18.3), thus reducing the likelihood of the slab being damaged by the work intended to stabilize it. It is important that the slab remains sufficiently intact to span between the pile heads. Serious damage must be avoided during pile driving; otherwise it will be necessary to break up and replace the slab – an unplanned expense. A slab less than 100mm thick (if only in places) cannot be guaranteed to span between pile heads, and would not be suitable for mini-piling. Weak slabs would be equally unsuitable. If in doubt, a simple *in situ* impact test can be carried out, using a Schmidt hammer or similar. There is no generally accepted 'minimum suitable strength' for a slab to be mini-piled, but a strength of below 20 N/mm^2 would raise doubts. The

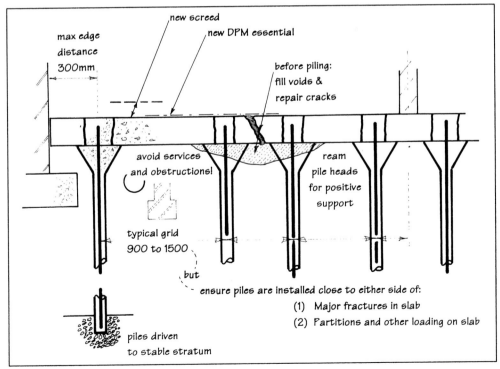

Figure 18.3 Mini-piling.

repairer might think it prudent in the face of such doubts to replace the existing slab with a new one. Mini-piling could still be used as the support.

The main limitation of mini-piles is their slenderness. Piles transfer loads in the same way as columns, compressing as they do so. As discussed in Chapter 2, elements in compression may fail by buckling if they are slender. Pile slenderness is defined as length, measured between points of lateral restraint, divided by diameter. A slenderness calculation based on the assumption that restraint is provided only at head and toe may be conservative, if some restraint is offered by the material (hardcore and soil) in between, even if it is loose. Whether it does so is a matter for individual judgement. There is no rule for all occasions. If the hardcore is stabilized by grout, slenderness may not need to be considered at all (but specialist advice should be taken). Without specialist advice, it would be best to define the piles' length as the distance from mid-slab to one metre below the top of stable stratum, and to restrict this length to forty times the pile diameter. Most mini-piles are steel cases driven to a set (Chapter 17), and then filled with concrete. There is no room to fix a cage of reinforcement, as would be normal for larger piles, and a central bar is usually implanted after concreting.

It is essential to place a new damp-proof membrane at top of slab level. This will mean that the new screed will be unbonded. It ought to be 65mm or more thick (Chapter 10), which could be an embarrassment if the previous screed was thinner. It is not ordinarily possible to change the ground floor level, even by a few millimetres, because door openings would have to altered. A possible solution would be to use a non-cementitious screed, if it can be laid within the available space. Asphalt, for example, would serve as membrane and screed.

Timber suspended floors

If local problems are caused by rot, it will be necessary to ensure that its causes are not allowed to strike again. Requirements for improving conditions can be more onerous than the structural repair itself. It is sometimes difficult, for example, to make the sub-floor ventilation effective. There needs to be a sufficient supply of air from outside (air bricks at 1500mm centres around the entire perimeter) without obstruction to cross-ventilation (all internal walls below floor level to be honeycombed, with no stale areas). If it is too late to provide these minimum standards at reasonable cost, it will pose awkward problems, because ventilation is as important as dry conditions. On such occasions, it is tempting to consider wholesale replacement with a ground-bearing concrete floor. That would be satisfactory from a structural point of view, provided that:

- the new floor can be provided with sufficient insulation to satisfy modern energy saving requirements; and either
- the soil has sufficient strength to support a ground-bearing slab, and is stable and inert; or
- a precast concrete floor can be specified that would not overload foundations.

Any damp that previously rose from the soil and evaporated up through the floorboards, or out through air bricks, would be held back by a new solid floor. It may well stay beneath the floor, harmless; but in some circumstances (combining differential vapour pressure and material porosity), it might be driven up the walls, appearing on the surface as rising damp. A solid floor solution may therefore have to be rejected, if rising damp is perceived as a possible unwelcome side effect that cannot, for whatever reason, be forestalled by installing damp-proof courses.

Concrete industrial slabs

The further beneath the surface the problem is, the more drastic the cure needs to be. The options of grouting and mini-piling are available for light industry, as they are for homes and offices; but these methods cope less easily with the demands of heavy industry. Grouting dissuades settlement under self-weight, by filling voids in the sub-base and subgrade, but it adds only modestly to the floor's capacity to support imposed loading. Although there is a piling solution for almost any situation, cost is in almost direct proportion to the imposed loading. There comes a point when repair would be more expensive than a total replacement of slab and sub-base including, if necessary, a slice of subgrade (soil). Any cost calculation should take into account the cost of interrupting use of the building, which would depend largely on the length of contract.

The other two generic problems, wear and corrosion, are not structural in origin, and will not be discussed in detail here. However, their separate demands will influence costs, and therefore options, for structural repair. In the case of wear, the quality of the top few millimetres of concrete is critical. It is generally found that, unless this layer has reasonable tensile strength (about $0.75N/mm^2$), it will wear badly. Optional remedies, if this is the only problem, include applying a hardener or replacing the weak layer with a resinous screed. Specialist advice is needed. The floor surface also has to resist chemicals spilled by accident, as part of the industrial process, or as part of cleaning. If the surface is not immune to any of these, corrosion will eventually produce an unsuitable surface. Again, specialist advice should be sought.

Chapter Nineteen

Timber

Good quality timber outlasts the economic life of the building that houses it, unless it is challenged by decay, overstress or gross movement of its support. This chapter discusses standard methods of assessing, repairing and strengthening timber.

Home-grown hardwoods were used in medieval timber frames (Table 27.1, *page 285*). Imported softwood is now used in most new buildings. Hardwood comes from broad-leaved deciduous trees, such as oak and elm. Softwood comes from coniferous trees, such as fir, pine, spruce and hemlock. Hardwood is usually (but not inevitably) more difficult to cut and plane and more durable than softwood. There are exceptions. The very easily worked balsa, for example, is a hardwood; and willow, although a hardwood, is less durable than most softwoods. Table 19.1 lists a few species in order of durability, strength and stiffness, with rough comparative values. Actual values are not given because there are many variations according to grade and condition. Moreover, working stresses would be reduced by varying degrees to account for individual usage (type of load, type of stress, slenderness, and so on).

Table 19.1   Timber properties

Species	Durability	Strength	Stiffness
Willow	Perishable	–	–
Sitka spruce	*Non-durable*	*2*	*4*
Douglas fir	*Moderately durable*	*2.5*	*5*
Oak	Durable	5	6
Greenheart	Very durable	10	10

Italic type indicates typical range for imported softwoods.

Basic procedure

Timber repair proceeds as follows:

- Provide temporary support if required.
- Remove the cause of damage if possible.
- Assess (current) strength in the light of (future) loading.
- Consider the non-structural constraints on structural repair, such as conservation and safety in fire.
- Design the repair.
- Cut out rotten or weak timber where necessary.
- Apply chemical treatment if appropriate.
- Carry out the structural repair.
- Make good, decorate and tidy up.
- Record what has been done.
- Monitor if necessary.
- Arrange to inspect at normal maintenance intervals.

Each case is unique. Sometimes every step is important; at other times, removing the cause of damage, confirming structural adequacy, and recording the event afterwards, is ample attention.

Temporary support

If temporary support is in place before the repairer first visits, it should be checked and adjusted if necessary, because damage can be caused by bad propping. All raking shores and vertical props should provide continuous safe load paths through to ground level. Elements of the building should be relied upon for support only when alternatives are impracticable, and then only with utmost care. A substantial timber floor beam, for example, may be a tempting platform for props; but even if it remains safe in this role, it may deflect more than previously, causing unnecessary damage to finishes. Clearly it would be better not to pass any load through sensitive finishes; where this cannot be avoided, the risk of damage can be reduced by using a large number of lightly loaded props, installed with minimal tightening. Generous spreaders with cushioning material at the contact face would also reduce the risk of damage.

Temporary work should be regularly checked to ensure that it remains fit.

Removing the cause of damage

The main causes of damage are decay, overstress and gross movement of the supporting structure. Decay is the most common. Eliminating its main cause – damp – is not always straightforward. There may be more than one source. Tracing all possible sources can sometimes be more elusive than finding the causes of structural damage. It is even possible to introduce another cause of damage through curing an existing damp problem completely (Figure 19.1). When there is even a slight risk that damp may persist, it is advisable to monitor conditions, and perhaps apply treatment to ensure survival during the period of uncertainty.

Overstressing may be the result of increased loading or the consequence of timber having been weakened by prolonged decay or unwise alteration. Sometimes stress can be reduced to a safe level by changing the load paths or repositioning the load (for example, locating a library on the ground floor instead of the first floor). Failing such a fortunate solution, there is no option but to strengthen and stiffen the distressed timber, or replace it.

If the damage is found to have been caused by movement of the supporting structure, then obviously this movement should be arrested (if it has not yet stopped of its own accord), before starting on the timber repairs. If there has been serious loss of bearing, this should be restored before any other permanent work is done.

Assessment

Timber's strength is reduced by saturation. It regains strength on drying out. During the drying-out period, it may benefit from temporary support or load reduction (such as removing stored materials or putting up barriers to restrict loading of the area supported by the weakened timber).

The tunnelling action of beetles reduces strength and stiffness to the extent that it removes wood. But what remains is sound. Serious weakness can occur in two areas: a labyrinth of passages near the face can destroy the value of the outermost parts of the member; and deathwatch and longhorn beetles can, if conditions are favourable to them, eat away large internal volumes. Obviously, edge loss is much easier to detect than internal cavities, and fortunately it is more common. A surface examination will usually be sufficient if all of the following are satisfied:

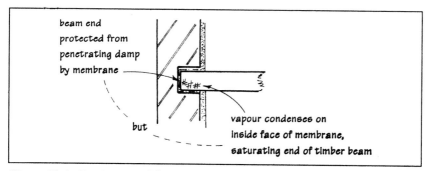

Figure 19.1 Persistence of damp.

- deathwatch and house longhorn can be excluded
- exit holes are sparse
- the timber members most affected can share or shed load without serious consequences.

Loss of strength or stiffness will be fairly modest in such a case. Where there is more uncertainty, assessment should not be attempted without further examination, especially if the timber most affected belongs to a principal structural member. In ascending order of usefulness, examination techniques include: sounding, drilling, coring, cutting inspection holes and load testing. Destructive investigation may not be welcomed by the owner and may not be permitted if the building is historic. Load testing is only valuable if it can be guaranteed that little or no further attack will occur.

Edge timber that has been ravaged by beetle attack can be removed by scraping (defrassing). This simultaneously establishes the extent of damaged material, removes it in preparation for repairs and improves the effectiveness of any treatment applied to the underlying timber. Scraping does, of course, change the appearance of the member, which may or may not be permitted. Deeper cavities can be detected by sounding, but can only be accurately defined by drilling or cutting from permitted access points.

When it comes to comparing a member's current strength with its future obligations, it is worth making sure we understand its true structural role. The following example returns to the theme of basing the structural model of the building on its performance (Chapter 4). Roof purlins are considered to be primary structural members, because they support rafter loads and transfer them to cross walls, struts or main trusses. In practice, however, the distance between purlin supports in many older roofs is so large that the purlins are not stiff enough to offer much more than token support to the rafters. They could be made strong enough to carry out their ideal structural function, fully supporting their intended share of imposed loads. There is little point in doing that, however, if the roof has performed gracefully for many years without the full benefit of the purlins. Nevertheless, even 'inadequate' purlins have value. They provide part of the roof's longitudinal stiffness, and they carry out a modest degree of redistribution, transferring a proportion of loading from the less competent rafters to their stiffer neighbours. These secondary functions increase the total strength of the roof, in some cases by as much as 25%. Depending on the merits of the case, the repair choice would lie between restoring the member to a secondary role and upgrading it to a primary one.

If a member has a genuine primary structural function, there is clearly no alternative to making it fit for purpose. Assessment starts with an estimate of characteristic strength. This requires judgement. New timber is usually ordered by species and grade. Existing timber can be identified by species (if not by visual inspection, then by sending away a small sample to be examined under the microscope). Grading is more difficult. It is, in fact, a specialist skill. Even a specialist has difficulty grading timber *in situ* because, for example, it is possible to estimate the detrimental effect of knots only if all four edges of the timber can be viewed. When information is limited, it is normally safe to assume the lowest grade for the species quoted in British Standards or orthodox texts. At this stage, the repairer has to decide whether to take into account the structural value of any non-structural or secondary elements that might be taking a share of the load (Chapter 2).

Figure 19.2 shows three examples of secondary elements adding to the strength and stiffness of the primary members.

Constraints on repair

The inescapable inaccuracies of structural analysis (Chapter 4) mean that we must be wary of the casual removal of apparently non-structural elements such as wall infills. Frustratingly, we must be equally cautious about assigning structural properties to them. Before placing reliance on a 'non-structural' element, the repairer needs to be satisfied that it:

- has adequate capacity
- is (or will be) adequately connected to its structural partner
- is unlikely to be dismantled in ignorance.

If these three criteria are satisfied, the two parts ('structural' and 'non-structural') can be said to be acting compositely. Each case needs to be judged carefully on its individual structural merits. In the case of Figure 19.2c, for example, floor joists should be assumed to be helping the main beam only if they are:

- tight within their housing
- entirely within the main beam's bending compression zone.

Only if the latter were true would the joists be capable of transferring force along the length of the main beam. Joists within the beam's tension zone (ceiling joists, for example) would not be able to transfer tension unless soundly connected to the main beam by resin, dowel or tensile strap.

The third criterion, avoiding accidental dismantling, is difficult to guarantee. It may be wise to carry out a risk assessment to explore the consequences of failing to satisfy it. As with the health and safety assessment (see Appendix A), risk may be calculated as *severity times likelihood*. Severity would depend largely on the consequences of structural failure: on whether, for example, damage would be widespread or local, disastrous or inconvenient. The likelihood would be judged on the possibility of dismantling, and on how rapidly this could trigger failure. If failure is then possible with only the dead load acting, likelihood would be increased. If failure is only possible once a large proportion of maximum imposed load was in place, likelihood would be reduced.

FIRE SAFETY

There is space here to discuss fire design only briefly. Its principles are:

- to provide safe escape
- to deter fire spread
- to provide means for fire fighting
- to prevent collapse while people are escaping and the fire is fought.

Proposals for repair should not compromise any existing safety measures or any that are about to be implemented.

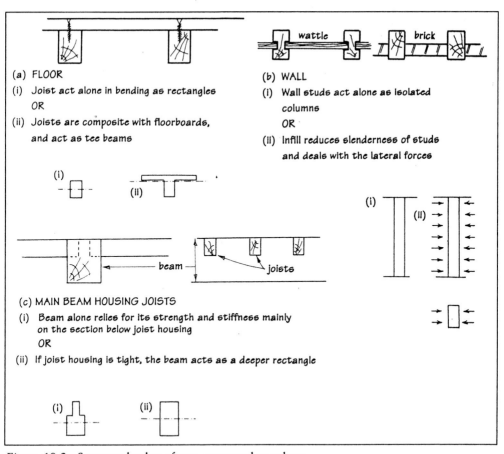

(a) FLOOR
(i) Joist act alone in bending as rectangles
     OR
(ii) Joists are composite with floorboards,
     and act as tee beams

(b) WALL
(i) Wall studs act alone as isolated
     columns
     OR
(ii) Infill reduces slenderness of studs
     and deals with the lateral forces

(c) MAIN BEAM HOUSING JOISTS
(i) Beam alone relies for its strength and stiffness mainly
     on the section below joist housing
     OR
(ii) If joist housing is tight, the beam acts as a deeper rectangle

Figure 19.2   Structural value of non-structural members.

Means of preventing collapse are measured by notional fire resistance. The Building Regulations give minimum periods for which the building must safely support its structural loading. These can be assured either by covering the structural elements (by boarding or casing, or applying intumescent paint) or, in the case of timber, by making them over large, so that they remain safe after they have been charred for the statutory period. (The charrable material is known as sacrificial timber.) Typical rates for charring are mentioned in Chapter 15. Specialist orthodox texts (such as *Timber Designers' Manual*) should be referred to for detailed design.

Precautions against fire can be defeated by inattention to detail, such as failing to make allowance for the additional charring surfaces at joints, where fire can penetrate deep into the structure. Steel bolts and plates should not be left unprotected. They quickly lose strength during fire and they rapidly conduct heat to the interior of timber and masonry. Coverings should not have gaps.

OTHER CONSTRAINTS

The rules of conservation (Chapter 5) can severely reduce repair options in historic buildings. So might task difficulties, in buildings of any age (Figure 19.3). It may be awkward to lift heavy flitch plates into position, especially if there is little room for lifting equipment, and more so if working space is limited by the building remaining in use during repair.

Chemical treatment

When chemical treatment is unavoidable, the first step is to check whether bats are roosting in the roof space, as they are protected by law and may not be disturbed or endangered by toxic fumes. Treatment may consist of:

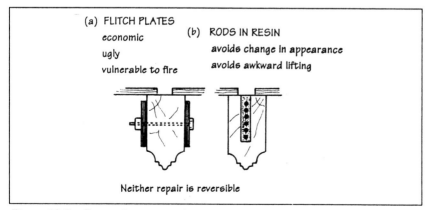

Figure 19.3   Constraints on repair.

- surface brushing or spraying
- drilling into the member and irrigating or injecting
- applying thick emulsion (usually referred to as 'mayonnaise') to the surface.

(Extra precautions may be necessary with dry rot or deathwatch beetle. Only repeated smoke treatment can halt a severe and persistent outbreak of the latter.)

The specification of treatment should be entrusted to an experienced surveyor.

CORROSION AND CONSEQUENTIAL DAMAGE

Sometimes previous treatment has itself been the cause of damage; the chemicals, probably combined with a higher than normal moisture content, may have led to corrosion of metal fasteners by electrolytic action (Chapter 15). Corroded fasteners should be replaced by material of equivalent quality and protected, if necessary, from a second round of corrosion. The timber itself may have been attacked and weakened, and may need local repair. It will also be necessary to check for, and remedy, any structural damage caused by the fastener's loss of performance. An illustration of the consequences of corrosion and allied problems can be drawn from the earliest trussed rafters. Figure 2.3c (*page 8*) outlines a typical trussed rafter for a domestic roof. Each joint is fastened by a gang-nailed gusset.

Few problems arise with today's products, but in the early days of manufacture and assembly it was not uncommon to discover gusset corrosion, with consequent rotation at the joints. Other problems were ill-fitted gussets, gussets damaged by careless lifting, out of plumb fixing, overloading at water tanks, and lack of bracing – although rarely all in the same building! Where a similar combination of defects is feared, a careful structural appraisal should precede action. That might include one or more of the following:

- temporary propping and possible jacking to eliminate the more serious distortions
- replacing corroded gussets (plywood is often a satisfactory alternative)
- local strengthening
- providing bracing to resolve out of plane forces (in some cases such forces can be large and plywood membranes are often the most economic solution)
- tying the roof in to external walls.

Similar corrosion problems, requiring more than material replacement, can arise with *in situ* girders or trusses or flitched beams.

Designing structural repair

There are three main choices for strengthening:

- replace with timber, like for like
- convert to composite
- replace with different material.

REPLACING LIKE FOR LIKE

Replacing like for like should be done with new timber of the same species as the existing member. Second-hand timber may be used if it is workable and clean – devoid of nails and other inserts. The moisture content of the introduced timber, whether new or old, should be as close as possible to its intended environment; although, however close it appears to be, some drying shrinkage and twisting will take place. (Figure 15.1, *page 133*, shows typical moisture content ranges. Green timber is usually appropriate for sole plates; dry timber in such a location would be liable to swell.) Connections should be chosen partly for their ability to survive moisture movement in the timber. For example, through bolts are usually better than screws (but screws may be preferred for other reasons). Screws are nearly always better than nails, which are loosened by shrinkage.

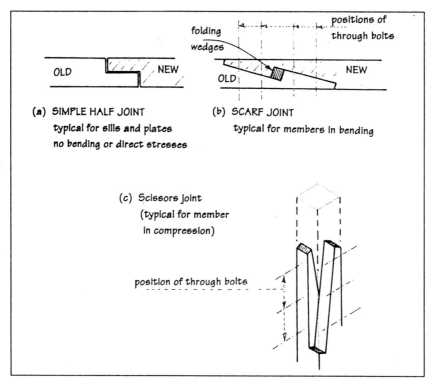

Figure 19.4   Jointing new and old.

Joints between existing and introduced timber must be designed for the expected stress. Cills can have simple half joints (Figure 19.4a) because their stresses are low. Members taking bending moments must be capable of maintaining internal equilibrium at the joint (Figure 19.4b). Compression members must form a tight fit (Figure 19.4c) to avoid unnecessary take-up (strain) as the load is introduced to it.

CONVERTING TO COMPOSITE

The simplest composite repair is the steel flitch plate (Figure 19.3a). Flitching is clearly easier and safer to do in the factory, or in a part of the site set aside as a workshop; but it is not often practicable to remove a large beam and replace it. *In situ* flitching can be inefficient if account is not taken of existing stress in the timber. The timber beam will already be stressed with the dead weight of itself and any floors, ceilings and posts it might be supporting; but the flitch plate will be unstressed when it is dropped into the slot and bolted into place. At that moment it will be supporting nothing. Only when the imposed (live) load begins to build up will the plate start to take any share of the load.

Suppose, for simple illustration, that the flitch plate has exactly the same stiffness as the timber beam, and suppose dead load is the same as imposed load. When the composite beam is fully loaded – to the maximum imposed load – the timber will be taking the full dead load and half the imposed load. It amounts, in this hypothetical case, to three-quarters of the total, with the steel plate taking only one-quarter. That is very inefficient, considering that the full imposed load is rarely achieved and, therefore, the expensive steel plate would usually be supporting only a small fraction of the load actually carried by the beam.

A better arrangement is illustrated in Figure 19.5 (*overleaf*). The timber beam is relieved of stress by placing a line of props or jacks beneath it and deflecting the beam upwards to the point where all the dead load is transferred into the props or jacks. (It is sometimes easier simply to isolate the beam from all its load except self-weight.) The flitch plate is inserted. Connections are made. The unstressed beam then becomes fully composite and, when the props are removed, the two materials will share all the dead and imposed loads in proportion to their stiffness (equally, in this example). This uses the steel more economically, and it reduces deflections.

As with all composite material, there must be sufficient connection to ensure that the load is shared as intended. If the steel plate takes 50% of total load, the bolts shown in Figure 19.5 must be able to resist a shear load of at least half the total dead and imposed load on the combined beam. (A slightly greater bolt capacity than the theoretical value would be useful, because in time the timber component will creep and transmit a

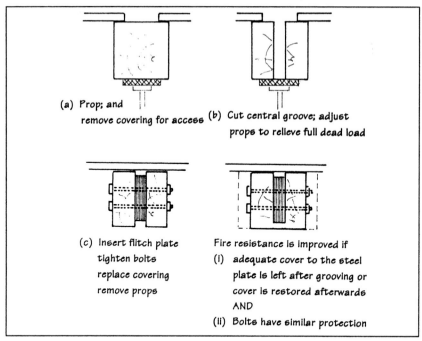

Figure 19.5 Installing simple flitch plate.

portion of its initial share of load to the steel.) Finally, if there is a fire in the space below, the original beam would char on three faces but the beam as flitched would char internally, because the steel would rapidly conduct heat into the interior. So will the bolts, if they are exposed. Furthermore, the two materials will act differently in fire. Timber, being a good insulator, does not weaken below the line of charring, whereas steel will steadily weaken over its entire section as its temperature rises. Final proposals would have to deal with these shortcomings.

REPLACING WITH DIFFERENT MATERIAL

Various grouts, moulds and thixotropic fillers have been developed for the *in situ* replacement of decayed wood, bonding wood to reinforcement and filling cracks. Most of these mouldable materials are based on epoxy resin, whose structural behaviour is compatible with timber. If a suitable composition is chosen, it can be sanded, planed or chiselled. As there are many different compositions, manufacturers' advice should be sought on selection and application, and on any safety precautions to be taken during the work. Manufacturers' advice should also be sought on fire resistance. The concept of charring would not apply to exposed epoxy resin.

When the whole element has to be removed – a rotten timber lintel is the most obvious example – propping and, if necessary, shoring should first be installed to maintain equilibrium until it is safe to transfer all loads onto the replacement. The replacement may be a modern proprietary lintel or *in situ* reinforced concrete. The latter can conveniently be the same size as the original timber in most cases.

Roof spread

Roof spread is one of the more visible structural problems with traditional buildings. The majority of spreads can be safely left alone. They might perhaps be monitored for the rate of continuing movement, as there is a tendency to creep, but drastic action is seldom needed. Some cases benefit from a more positive eaves connection. Figure 19.6 (*overleaf*) shows a necessarily fussy solution chosen to stall severe roof spread after it had been rejuvenated by careless renovation. The total horizontal movement had become worrying, but the owner requested a solution that did not entail any rebuilding or obvious alteration, such as an inserted frame.

Typical details

Replacing the rotten beam end is one of the more common timber repairs (Figure 19.7, *overleaf*). Preventing further damp is also essential, to avoid a reappearance of the rot. However, there may be no guaranteed method of doing this if the barrier between the end of the beam and the outside face of the building is a thin section of porous masonry. It may be better not to use a solution involving built-in timber in such cases. Alternatively, an impermeable barrier may be placed in the gap between beam end and masonry. If such a barrier is used, the repairer must be satisfied that penetrating damp cannot bypass the barrier and, equally important, the barrier itself is unlikely to encourage interstitial condensation, which could saturate the beam end.

Figure 19.8 (*page 193*) shows some typical beam end repairs. Figure 19.9 (*page 194*) shows options for improving the bending strength within the span of the beam. Tensile capacity is most easily supplied by stainless steel or timber straps (Figure 19.10, *page 194*). If members are large, it is possible to transfer tension entirely through the faces of the joints (Figure 19.11, *page 195*), without introducing a new element. When a main tie is removed from a timber frame and cannot be replaced, the best option is usually to provide a new frame to 'bend' the tie force around its original direct route (Figure 19.12, *page 195*).

Splits in timber can create serious weakness, especially in bending. Some of the methods shown on Figure 19.9 may be used to compensate

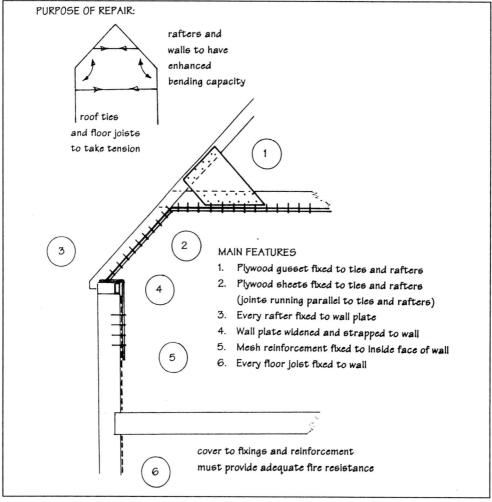

PURPOSE OF REPAIR:

rafters and
walls to have
enhanced
bending capacity

roof ties
and floor joists
to take tension

MAIN FEATURES

1. Plywood gusset fixed to ties and rafters
2. Plywood sheets fixed to ties and rafters
   (joints running parallel to ties and rafters)
3. Every rafter fixed to wall plate
4. Wall plate widened and strapped to wall
5. Mesh reinforcement fixed to inside face of wall
6. Every floor joist fixed to wall

cover to fixings and reinforcement
must provide adequate fire resistance

Figure 19.6   Arresting severe roof spread.

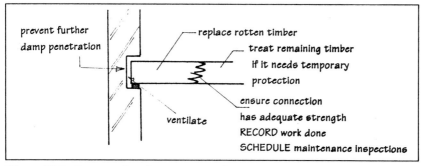

prevent further
damp penetration

replace rotten timber

treat remaining timber
if it needs temporary
protection

ensure connection
has adequate strength
RECORD work done
SCHEDULE maintenance inspections

ventilate

Figure 19.7   Beam end replacement: requirements.

for such loss of strength, but in many cases simple stitching or injection by epoxy resin will achieve the same result for less cost (Figure 19.13, *page 195*). (Old splits can usually be left alone if the member is performing adequately and there are no plans to increase loading.)

Bonding timber (Figure 15.7, *page 143*) that is tight to the masonry, in good condition, and highly unlikely to deteriorate can be left undisturbed. Otherwise it should be removed. This must be done piecemeal (Figure 19.14, *page 196*); otherwise, whole wall panels will be severely

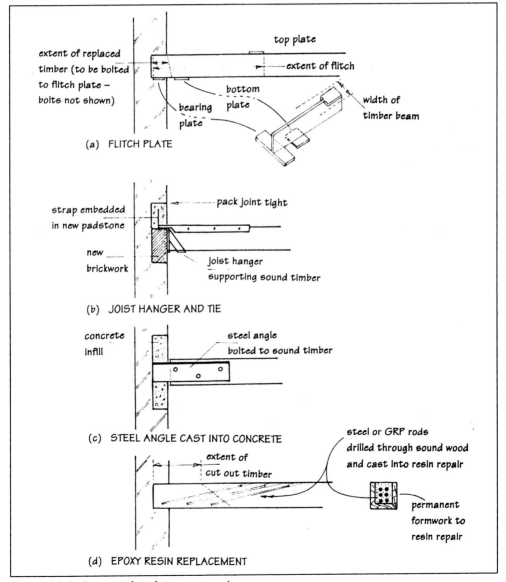

Figure 19.8 Beam end replacement: options.

reduced in strength during the work, possibly with embarrassing consequences. If the bonding timber is not really serving any structural function (beyond the straightforward passing of loads vertically downwards),
it can be replaced by brickwork, with slate packed into place to prevent
movement while the mortar or resin sets (Figure 19.14b). If the bonding timber passes above or below large openings, it may be beneficial to
retain the continuity it was intended to provide (Figure 19.14c).

Timber decay is often worst at joints, and the choice of repair usually
lies between splicing in new timber or supplying substitute material such
as epoxy resin (Figure 19.15, *page 196*). The latter alternative destroys
the flexibility of the original joint and will certainly not be a reversible
repair, aspects which may or may not matter. Each repaired joint will have
to be strong enough to match the stress created by its acquired stiffness,
like a modern frame. This is seldom a serious problem, but it is a point
to be checked.

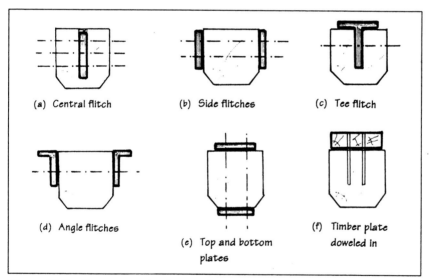

Figure 19.9   Beam strengthening: options.

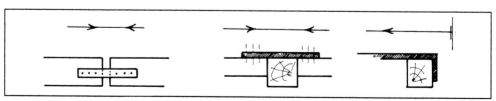

Figure 19.10   Tension straps.

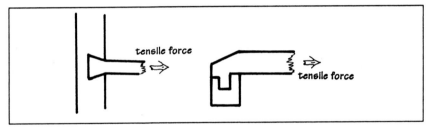

Figure 19.11   Tensile joints.

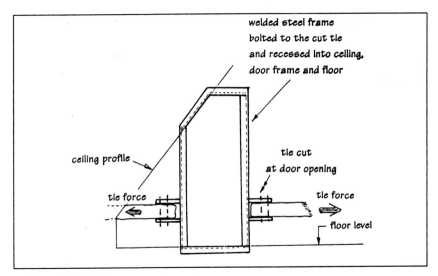

Figure 19.12   Frame round opening.

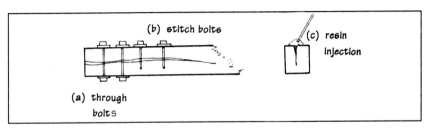

Figure 19.13   Repairing splits.

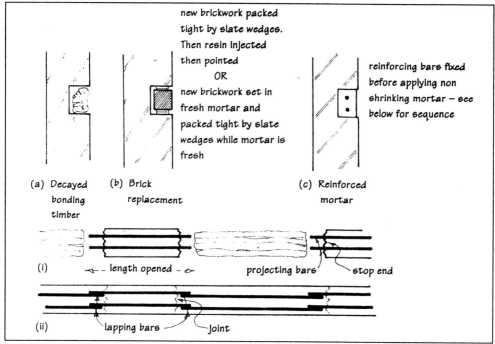

Figure 19.14   Replacing bonding timber.

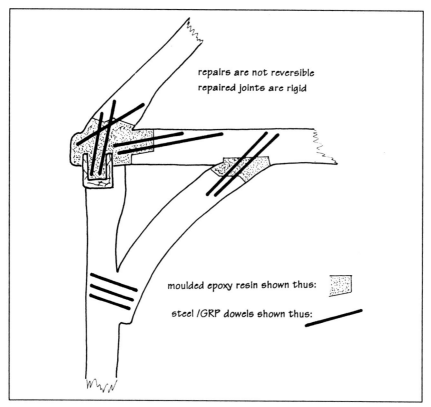

Figure 19.15   Restoring connections.

Chapter Twenty

Masonry

In most cases, masonry enjoys a generous margin of safety between its strength and the level of stress it must sustain. Unfortunately for the investigator, the transition from safe to unsafe can sometimes be brief and lacking in symptoms. After discussion of assessing masonry, this chapter describes solutions to specific and general problems. Quality of workmanship is important to the success of masonry repair; in a few cases where it is most critical, a recommended method of working is included.

Chapter 2 discussed the difference between ductile and brittle materials (Figure 2.11, *page 16*). Ductile materials give warning of failure, as their strain becomes disproportionately great at its approach; brittle materials fail without warning. The concept can be borrowed to discuss the behaviour of damaged buildings. In damaged buildings, the extent to which failure is preceded by warning symptoms depends not only on the ductility of the materials, but on many factors, including the type of stress and how the materials are fixed into the building. To avoid misusing terminology, we should label the extremes of building behaviour as 'gradual' and 'sudden', rather then 'ductile' and 'brittle' (Figure 20.1, *overleaf*). In the case of masonry, gradual failure is typified by walls bending in their own plane as a result of settlement or subsidence. Sudden failure is typified by walls leaning or bulging out of plumb as a result of eccentric loading.

In-plane movement and damage

In 1981, the Building Research Establishment (BRE) produced their Digest Number 251: *Assessment of damage in low-rise buildings* (since revised). This provided for the first time a visual classification of masonry damage, appropriate for low-rise masonry buildings, which are more likely to be affected by in-plane bending than by shear force. (Shear force is a more common cause of damage in taller buildings.) The digest defines three broad levels of damage:

- *aesthetic:* affecting only the appearance
- *serviceability:* affecting some aspect of performance
- *stability:* threatening structural integrity.

These break down into six categories (aesthetic: categories 0,1,2; serviceability: 3; stability: 4 and 5). Table 1 of Digest 251 is reproduced here as Table 20.1.

When damage is category 3 or worse, foundation movement is the most likely cause (statistically, for low-rise buildings); below category 3, foundation movement is one of several possible causes. Diagnosis (Chapter 7) is more difficult for slight damage than for severe damage.

Out-of-plane movement and damage

BRE Digest 251 classification should not be used blindly. Very occasionally, failure is sudden, without pausing within the lower categories. These failures usually arise from leaning or bulging, which are difficult to judge by eye as they approach the point of disequilibrium. In older buildings, such distortions sometimes accrue slowly and may cause rupture at junctions with cross walls or minor cracking and bulges; but these useful symptoms are often covered up in the course of maintenance, leaving no clue on view, so that only instruments can reliably detect the growing problem.

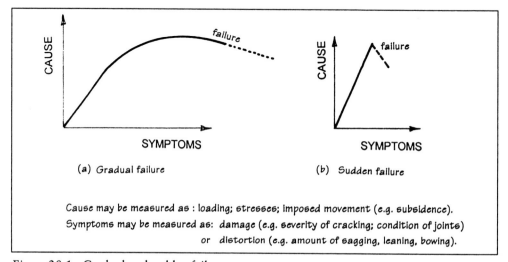

Figure 20.1   Gradual and sudden failure

Table 20.1† Classification of visible damage to walls with particular reference to ease of repair of plaster and brickwork or masonry

Crack width is one factor in assessing category of damage and should not be used on its own as a direct measure of it.

Category of damage	Description of typical damage *Ease of repair in italic type*
0	Hairline cracks of less than about 0.1mm which are classed as negligible. *No action required.*
1	Fine cracks which *can be treated easily during normal decoration.* Damage generally restricted to internal wall finishes; cracks rarely visible in external brickwork. Typical crack widths up to 1mm.
2	*Cracks easily filled. Recurrent cracks can be masked by suitable linings.* Cracks not necessarily visible externally; *some external repointing may be required to ensure weather-tightness.* Doors and windows may stick slightly and *require easing and adjusting.* Typical crack widths up to 5mm.
3	Cracks which *require some opening up and can be patched by a mason. Repointing of external brickwork and possibly a small amount of brickwork to be replaced.* Doors and windows sticking. Service pipes may fracture. Weathertightness often impaired. Typical crack widths are 5 to 15mm, or several of, say, 3mm.
4	Extensive damage which requires *breaking-out and replacing sections of walls,* especially over doors and windows. Windows and door frames distorted, floor sloping noticeably.* Walls leaning or bulging noticeably,* some loss of bearing in beams. Service pipes disrupted. Typical crack widths are 15 to 25mm, but also depends on number of cracks.
5	Structural damage which *requires a major repair job, involving partial or complete re-building.* Beams lose bearing, walls lean badly and require shoring. Windows broken with distortion. Danger of instability. Typical crack widths are greater than 25mm, but depends on number of cracks.

*Local deviation of slope, from the horizontal or vertical, of more than 1/100 will normally be clearly visible. Overall deviations in excess of 1/150 are undesirable.

†Table 1, BRE Digest 251, 1995 revision, reproduced by permission of BRE.

Assessing out of plane distortion

CIRIA Report 111 (published by the Construction Industry Research and Information Association) provides the following indicators of stability:

- Where the wall is not well restrained, it retains a precarious equilibrium with leans or bulges of up to 85% of wall thickness (assuming a solid wall), provided it supports no loading from an upper storey or beam.
- With adequate restraint, the wall will support a concentric load while still retaining a precarious equilibrium at leans or bulges of up to 85% of wall thickness.
- An unrestrained wall supporting a concentric load reaches precarious equilibrium at 50% of wall thickness.

The CIRIA report contains clear illustrations and explanations of these principles. The information provides comfort for many investigators. It also inspires ideas for strengthening walls that are poorly restrained.

Figure 20.2 summarizes the symptoms that can be judged either visually or by verticality measurements. These provide a clue to severity (by classification of in-plane damage or out of plane movement), but also act as indicators of cause (Chapter 7). The pattern of cracking may imply that bending is in sagging mode (tension at foundation level) or hogging mode (tension at eaves level). This may be used as part of the evidence of cause, and will certainly be useful information when it comes to specifying repair. Openings alter crack patterns and make this judgement more difficult. Verticality measurements should be taken at close centres (never more than 300mm; less if necessary) if their pattern is to be used as an indicator.

Tall heavy walls

The benefits of taking verticality readings at close centres – both as an indicator of cause and as an aid to repair – can be appreciated by considering church walls and similar tall, heavy walls, which are often found to be leaning outwards. Sometimes the cause is roof spread; sometimes it is erosion or softening of the soil immediately beneath the foundations, if they are shallow (as they frequently are). Roof spread tends to create a concave distortion of the wall, viewed externally. This looks similar to Figure 20.2j if the masonry is weak, but closer to Figure 20.2l if it is strong. It may even approach a 'straight-line lean' if it is both very strong in itself and stiff compared with the soil supporting it. Erosion or softening of the soil creates a convex profile, viewed externally, akin to Figure 20.2k. This is because the softening or erosion tends to occur at the outside edge of the foundations, the area most vulnerable to leakage and

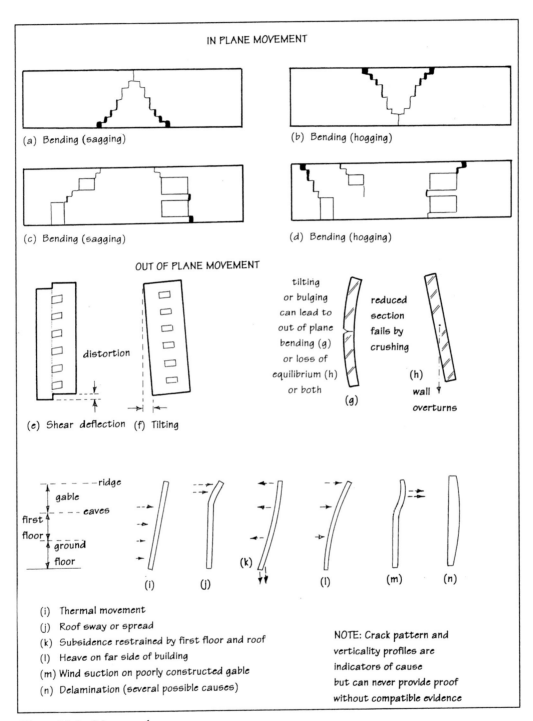

IN PLANE MOVEMENT

(a) Bending (sagging)

(b) Bending (hogging)

(c) Bending (sagging)

(d) Bending (hogging)

OUT OF PLANE MOVEMENT

distortion

tilting
or bulging
can lead to
out of plane
bending (g)
or loss of
equilibrium (h)
or both

reduced
section
fails by
crushing

(g)

(h)

wall

overturns

(e) Shear deflection   (f) Tilting

ridge

gable

eaves

first
floor

ground
floor

(i)

(j)

(k)

(l)

(m)

(n)

(i)   Thermal movement
(j)   Roof sway or spread
(k) Subsidence restrained by first floor and roof
(l)   Heave on far side of building
(m) Wind suction on poorly constructed gable
(n) Delamination (several possible causes)

NOTE: Crack pattern and
verticality profiles are
indicators of cause
but can never provide proof
without compatible evidence

Figure 20.2   Masonry damage.

storm water infiltration, and the lean originates at that level. As always, several indicators should be used before making a detailed diagnosis (Chapter 7). It is not unknown for both roof spread and soil movement to occur, simultaneously or in succession.

If the cause is active, it should be eliminated, by resolving the horizontal thrust of roof spread or protecting the soil from erosion or water content fluctuation. Then the repair specification should take into account, as a determining factor, the detailed measurement of distortions. (Serious leans or bulges will have reduced the wall's safety margin, which might now be most easily restored by buttressing or some other means of improving restraint.) Methods are discussed later in this chapter. If the cause cannot be removed, or the wall's margin of safety has vanished, extreme measures may be required, such as rebuilding, jacking to reduce load eccentricity (Chapter 17), or underpinning.

Analysis of masonry strength

The strength of both brick and stone can vary from source to source by a factor of more than 100, and it is not possible to grade the material by eye. Workmanship and deterioration serve to widen this variation still further (as discussed in Chapter 15, under *Masonry*). Except for identifiable products in good condition, strength can only be established accurately by testing.

There are, however, drawbacks to testing. Non-destructive testing (for example, using a rebound hammer) does not give reliable results. Destructive testing changes appearance, which may be unwelcome, whether or not conservation rules apply. The simplest means of testing is to drill out a core and crush it in a laboratory. There are pull-out tests for weaker material, including the mortar. *In situ* testing on strong masonry usually involves isolating a small panel or pier by cutting it free from the rest of the wall and applying a load above it. Unfortunately, several tests would be needed to obtain a reliable average.

In many cases, a precise value for strength is not necessary, provided it can be shown that the proposed repairs would raise the wall's margin of safety to above its level at the time of construction – assuming, of course, that it subsequently stood the test of time and was not built as an unsafe structure. The comparison between original and proposed is an engineering decision. It should take into account slenderness, load eccentricity, deterioration, and a careful review of initial, current and future loading.

Solutions

Some solutions are specific to the cause of damage. These are dealt with first; the heading of each section gives a strong clue to the cause under consideration. Typical repairs, independent of cause, will be dealt with later in this chapter.

Deterioration

Before structural repair can be specified, a decision has to be made about whether to reduce the rate of deterioration. Methods of doing so may include surface cleaning, the removal of salts, killing biological agents and, where appearance must be restored, *in situ* moulded repair based on mortar with carefully chosen fillers. These are specialist tasks, not to be implemented by the inexperienced.

Taking into account the future rate of deterioration, reduced or not, the choice then lies between doing nothing, rebuilding (at least replacing the most damaged elements), or compensating for any weakness by improving restraint. Chapter 22 illustrates typical restraint details for masonry.

Sulfate attack above ground

Sulfate attack on stone rarely needs attention for structural reasons; when it does, it is a specialist task. Sulfate attack on mortar is a more common problem. When the sulfates originate from the bricks themselves, from sea-dredged aggregate, or from acid rain, the only way to arrest the damage is to provide a barrier to the essential supply of water. Thin barriers, such as silicon, are unreliable and short lived. They can also introduce fresh problems, especially in old buildings, by interrupting the outward passage of water and vapour (Chapter 15). The best barriers provide an obstruction to driven rain, with an air and drainage gap between themselves and the surface they are protecting. The best materials for this are tiling and cladding. Obviously, they completely alter the building's appearance. Otherwise the choice lies, as usual, between accepting the continuation of the damage or rebuilding the wall. Sometimes the cost-effective solution is to rebuild exposed parts such as parapets and chimneys, and to repoint other areas, accepting that further attention may become necessary.

Chimneys may create sulfate problems in the form of hygroscopic salts, initially deposited within unlined flues and subsequently migrating towards the inside face of the brickwork. One alternative to rebuilding, provided the structure is sound, is to insert a stainless steel liner, assuming it will perform satisfactorily as a flue. Reinforcement can be fixed to the outside face of the liner, and the gap between it and the brickwork can then be filled with sulfate resisting concrete. The rate of filling must be slow enough to avoid the wet concrete bursting the brickwork.

Cavity ties

Any contractor can install replacement wall ties, but the task is carried out mainly by specialists, using a variety of fixings developed for the range of circumstances found in practice. Figure 20.3 shows the basic model, together with a standard procedure for investigating and correcting the problem. Step 6 of the procedure, estimating the future life of existing ties, can be fulfilled by sampling the metal and sending it away for testing, or by judging visually the surface condition of the ties. In the absence of corrosion, ten years of life can be assumed. Pitting of zinc covering and

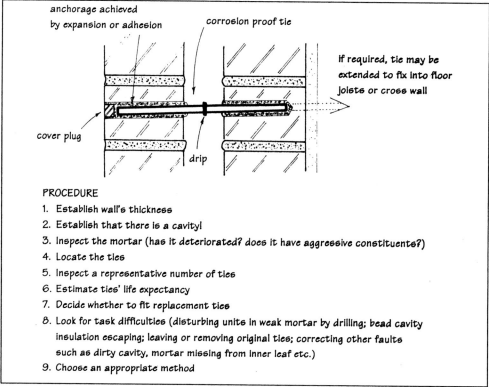

PROCEDURE
1. Establish wall's thickness
2. Establish that there is a cavity!
3. Inspect the mortar (has it deteriorated? does it have aggressive constituents?)
4. Locate the ties
5. Inspect a representative number of ties
6. Estimate ties' life expectancy
7. Decide whether to fit replacement ties
8. Look for task difficulties (disturbing units in weak mortar by drilling; bead cavity insulation escaping; leaving or removing original ties; correcting other faults such as dirty cavity, mortar missing from inner leaf etc.)
9. Choose an appropriate method

Figure 20.3   Replacement wall tie.

signs of white rust (but no more than a few spots of red rust) would suggest at least five years of life. Widespread red rust implies no useful life.

Wall tie failure would, in practice, take place over time, with gradual loss of wall strength, culminating in the two leaves acting independently (Figure 15.5, *page 140*).

Overstress

It has been known for wall tie failure to lead to the outside leaf peeling off in the wind as a result of increased slenderness (Figure 15.5). A heavy point load can crack the masonry immediately beneath it (Figure 20.4). This is one defect that should not be allowed to fester. The obvious answer is to provide a pad or spreader to reduce masonry stresses to acceptable levels. As previously noted (Chapter 15, *Unexpected weakness*), overstress can be the result of local faults such as poor quality bricks or slender piers.

New loads of any significance should be seated on pads or spreaders. Some thought should also be given to the effect of the extra loading on the foundations and soil. By definition, increased loading causes increased bearing pressure. An engineering judgement must be made as to whether the building and soil can distribute the extra load without causing damage, or whether the foundations need to be stiffened, widened or deepened.

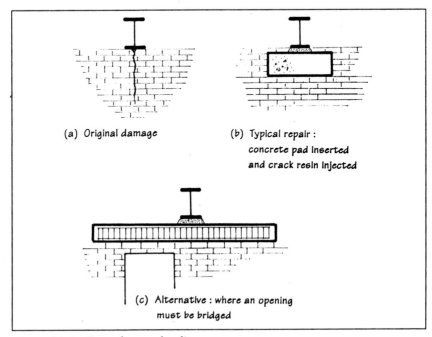

(a) Original damage

(b) Typical repair :
concrete pad inserted
and crack resin injected

(c) Alternative : where an opening
must be bridged

Figure 20.4 Cures for overloading.

Roof spread

Roof spread is a common defect in pitched roofs that are not tied at eaves level. Many cases can be allowed to creep gently for the rest of their lives. Sometimes, a small but useful degree of stiffening can be achieved by anchoring the eaves plate to the wall by means of a holding down strap. Chapter 19 includes a discussion, and an example of a fairly drastic repair, intended to arrest progressive movement (Figure 19.6, *page 192*).

Flat arches

Modern buildings usually have lintels above openings. Older buildings often relied on low-rise, or flat, arches, the latter often composed of brick-on-edge (soldier) courses. As long as the brickwork and mortar remain in good condition, this arrangement is satisfactory because it is possible for the masonry to develop an arch-type thrust line to support any significant dead and imposed loading. If, through alteration or deterioration, the thrust line is interrupted (Figure 20.5), the flat arch may fail. It may not be possible to restore arch action if, for example, it is the abutment that has failed. In this case, it may be best to convert the element to a beam. The failed arch can be supported on a modern lintel, or rebuilt using reinforcement within the new brickwork (Figure 20.6). Alternatively, if there is reluctance to alter appearances, it may be possible to reinforce the existing brickwork *in situ*. This is discussed later in this chapter, under *Masonry reinforcement*.

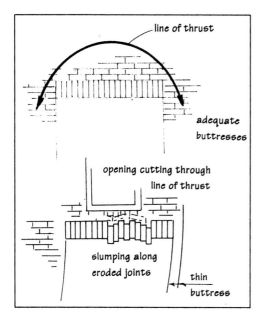

Figure 20.5
Brick lintel defects.

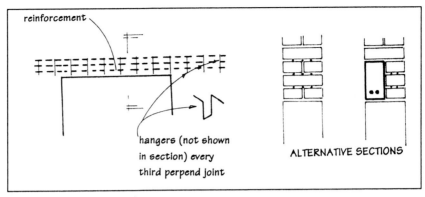

Figure 20.6   Reinforced masonry.

Loss of restraint

Mention has been made of the effect of slenderness on masonry strength. Modern regulations and standards encourage the restraint of walls along horizontal and vertical lines, so that slenderness ratios are economically low. Older buildings often lacked formal restraint, but enjoyed some accidental benefit simply from frictional contact between elements. This can disappear with time, unfortunately, leaving the building secretly weaker. The insertion of restraint will usually more than compensate for loss of the original accidental benefit. Examples are given in Chapter 22. Restraint forces are not usually high, but they must not be underestimated (Figure 20.7).

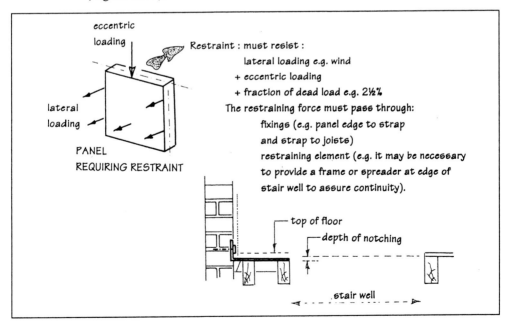

Figure 20.7   Restraint: principles.

Buttressing

Some buildings, such as churches, do not have floors and cross walls at close enough centres to provide suitable restraint. Their external walls were often built with buttresses to keep slenderness within safe limits. Sometimes, especially in some 'younger' churches, buttressing was not used, and a high slenderness ratio is part of the current problem. In such cases, buttressing may now appear to be the logical solution, especially for outward-leaning walls, where something positive is needed to arrest active movement. It is not a straightforward repair.

A new buttress has to settle under its own weight. In doing so, it some-times imposes its movement on the wall whose movement it was intended to restrict. Alternatively (Figure 20.8), it may tilt away from the wall as it settles. The cause of tilting is as follows. Soil immediately adjacent to the main wall will have previously consolidated under that wall's weight. The corridor of consolidation is usually narrow. The soil a short distance away from the wall face will always have been under negligible pressure; hence it will be significantly less consolidated, and therefore less stiff

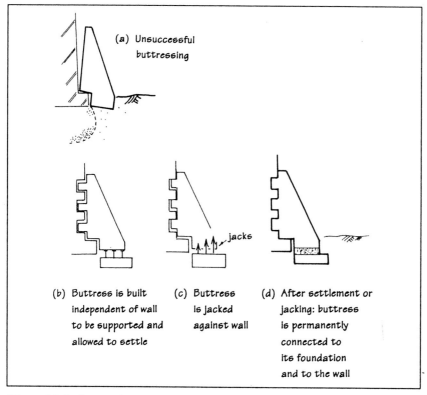

(a) Unsuccessful
    buttressing

jacks

(b) Buttress is built
    independent of wall
    to be supported and
    allowed to settle

(c) Buttress
    is jacked
    against wall

(d) After settlement or
    jacking: buttress
    is permanently
    connected to
    its foundation
    and to the wall

Figure 20.8   Buttressing.

(Figure 9.17, *page 102*). The difference in soil stiffness encourages differential settlement. As a result, the buttress departs from the wall, instead of supporting it. The buttress could, in fact, be reconnected to the main wall at the end of its settlement, and start performing its intended function. Unfortunately, its early performance often robs it of credibility and, as is commonly observed, owners are inclined to leave a tilted buttress as a freestanding monument, structurally independent.

It is better to let a new buttress settle before connecting it into the main wall (Figure 20.8b). If it can then be jacked into position (Figure 20.8c), it will be the more effective in deterring further distortion of the main wall. (If it is simply connected, without the modest prestress afforded by jacking, it will inevitably rotate slightly as it engages with the disturbing forces it has been designed to resist, and this will take a little of the edge off its success.)

Gross movement

Gross movement of the type illustrated in Figure 12.3 (*page 119*) requires a substantial designed repair to resist or accommodate it. If resisting it is not practicable and accommodating it is not safe, the only option is to rebuild.

Fire

The problems caused by fire are discussed in Chapter 13. The most common types of damage to masonry are: face spalling, material weakening, thermal cracking, and chemical change.

Face spalling and material weakening can be cured either by replacing damaged units or by applying render. These are simple 'typical' repairs, which are described later in this chapter. It is usually easy to tell the difference between sound and unsound masonry by hammer rebound, either informal or formal; if there is doubt, laboratory testing on a representative sample should resolve it. In many cases, weakening is found to be local and tolerable (the wall is still strong enough to do its job), in which case it is not necessary to cut out unspalled material.

Thermal cracking is caused by temporary stresses. When the cause has gone, a simple 'typical' repair is satisfactory in most cases. After severe fires, the distortions caused by expansion do not fully recover, and it is advisable to measure vertical profiles before deciding whether any parts need additional restraint or rebuilding.

Limestone, calcareous sandstone and lime mortar change to quicklime at 900°C. Any material that has gone through this change will need to be replaced.

Typical strengthening details

The simple examples shown in the rest of this chapter are independent of cause. It is assumed in all cases that the cause has been removed, and secondary effects such as leans, bulges and disruptions have been nullified.

Repointing

This is the most common, and most commonly abused, masonry repair. It is the normal remedy for mortar cracking and deterioration, but if it is not done properly, it can afflict the wall with permanent scars – and can even accelerate the deterioration it may have been intended to postpone.
Successful repointing should:

- have adequate strength
- have adequate durability
- be compatible with the existing mortar
- be a sensible compromise if it is difficult to satisfy simultaneously the other three requirements.

Table 20.2   Typical mixes for repointing

| | Conditions | |
	Mild	Severe*
Below ground	1:½:4½ or 1:0:4½	1:0:4
External: DPC to eaves	1:2:9 or 1:0:6 or 0:2:5	1:½:4 or 1:0:4
Chimneys and parapets	1:½:4½ or 1:0:4½	1:0:3 or 1:½:3
Internal	1:2:9 or 1:0:6 or 0:2:5	

Notes: Proportions are cement:lime:aggregate by weight. The above typical ranges are not exclusive. Selection depends on the required strength and durability, and on appearances and compatibility with existing material. A compromise sometimes has to be specified.
*In sulfate or acid conditions special cement may be needed.

Experienced operatives suppress the instinct to produce strong hard repointing. The wall gains no strength advantage if the repointing is stronger than the masonry units. It may be necessary to make the repointing stronger than the original mortar, if the latter has proved to be weak or non-durable, but even that decision should be taken warily. There are two major drawbacks to strong repointing:

- Mortar strength is usually increased at the expense of permeability. If the repointing is less permeable than either the units or the original mortar, it may cause damage by interrupting natural drainage and evaporation (Figure 15.6, *page 142*).
- If the repointing is stronger than the masonry unit, any damage arising from movement is encouraged to crack the unit in preference to the joints. (The opposite is preferred for appearance and ease of repair.)

Table 20.2 lists typical mixes. Materials should be like for like where possible. This may be a requirement if conservation rules apply. Sand:lime mortar should be replaced with sand:lime, although a small amount of cement can usually be added without disadvantage, where a compromise is needed on the grounds of either strength or durability.

To minimize the risk of shrinkage, the aggregate should be sharp sand and no more water should be included than is necessary to achieve workability. Ideally, the sand should be washed on site to remove all traces of dust and extraneous material.

Unless the whole building is being repointed, it is usually necessary to match the existing mortar colour. This can be done by choosing pigmented cement or appropriately coloured sand, or adding small amounts of additives such as brick or stone dust. Where appearance is vital, several trial mixes should be made before work begins. The new colour should, if anything, be slightly lighter than the original, as it will darken with age.

In preparation for repointing, the existing mortar should be raked out squarely to a minimum depth of 15mm. Where deterioration is severe, or cracking is more than hairline, the raking should be deeper. If it is necessary to deter inadvertent movement during this operation, horizontal gaps thus formed should immediately be packed with slate set back 15mm from the face. Removing decayed mortar is usually easy. Hard mortar can be difficult to chisel away, but the operative should resist the temptation to chase it out mechanically, as this may damage edges of the masonry units and widen the joint. If it is difficult to rake out without damage, the task learning curve should be located in the least conspicuous area of wall, where mistakes will not advertise themselves too boldly.

Once the raking is complete, including packing if necessary, the gap should be brushed or blown free of debris and dampened to encourage a bond between the new mortar and the existing unit and mortar. (A bone-dry surface would draw too much water from the fresh mortar, thereby creating a thin layer weakened by incomplete hydration; a soaking wet surface would draw no water at all, and may even lose water to the mortar, creating weakness by too much free water in the mix. Judgement comes with experience. Correct water content is more important for repointing than for new work, and experienced operatives should be employed.) New mortar should be ironed into the joints to fill all voids and provide a compact, well-bonded product, resistant to direct rain penetration.

Repointing should not be done in hot weather or when there is a risk of freezing. Rapid drying should be prevented by covering the work with tarpaulin or sacking, and dampening from time to time.

When it comes to choosing the bed joint profile, compatibility with remaining unpointed areas, if any, must be considered. Figure 20.9 shows some options. For the sake of durability (when it can be given the importance it deserves), it is best to avoid leaving ledges where water can collect and freeze. It is unfortunately easy to end up with a joint profile like Figure 20.9f, if an inexperienced bricklayer is employed to do the work, especially if the masonry units have some spalled edges, making it difficult to conform to one of the better profiles. This is definitely a profile to be avoided at all costs.

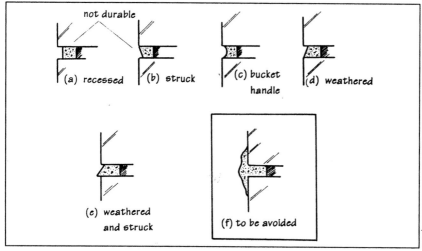

Figure 20.9 Repointing profiles.

Replacing masonry units

Units damaged beyond rescue should be carefully cut out from their surroundings and replaced by new units in fresh mortar, whose constituents should be carefully specified, as explained above. In some cases, especially with stonework, the new units should be carefully lifted and wedged into place, dampened and grouted. The grout should be stopped 20mm to 30mm back from the face, to allow for later repointing. New units should be like for like if at all possible, but at least compatible in technical performance. In the case of stone, the risk of a clash between old and new causing accelerated weathering must be avoided at all costs (Chapter 15). With brickwork, it is sometimes possible to take bricks from the inside face of the wall (or from a nearby free-standing wall whose weathering has been similar), or simply to turn the existing units round, to maintain like for like.

Masonry reinforcement

Reinforcement can be used for:

- stitching cracks
- providing an alternative route for loads, by creating beam action within the masonry
- strapping elements together.

Figure 20.10 (*overleaf*) shows two examples of reinforcing a solid wall. (The two leaves of cavity walls would normally need to be reinforced separately.) The reinforcement consists of small-diameter stainless steel rods, usually indented to improve bond, grouted into position using a suitable non-shrinking cementitious or resinous grout, and the face of the joint repointed (or the drill hole plugged).

Figure 20.11 (*overleaf*) shows a flat arch being converted to a reinforced brick beam by the insertion of rods into the lowest horizontal bed courses, endowing them with bending tensile strength. The individual soldiers should be prevented from dropping by anchoring into every third vertical joint between soldiers, taking care to avoid clashing with the horizontal reinforcement. The perpend joints between all units may need to be improved by repointing.

A similar detail can be produced for failed lintels or for any other situation where there is at least local advantage in converting plain masonry into a beam capable of developing bending tension. For example, if a proposed additional load above an existing opening is too much for the existing lintel, brick reinforcement may be a more discreet way of dealing with it than replacing the lintel.

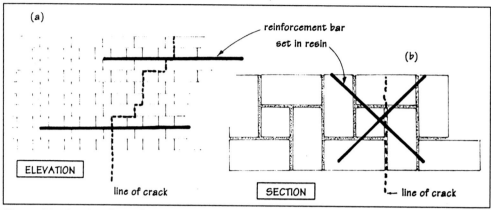

Figure 20.10   Crack stitching.

In cases where it is uneconomic to eliminate the cause of damage (Chapter 16), reinforcement may be a useful mitigator. (This only applies to in-plane movement. Out of plane movement can be mitigated only by restraint.) Sagging (Figure 20.2a,c, *page 201*) causes tension at foundation level, and a foundation strap (Figure 17.4, *page 160*) may be the best answer. Hogging (Figure 20.2b,d) causes tension at eaves level, and a continuous band of reinforcement just below the roof will reduce the unsightliness of consequential cracking. The detail is hampered if there is no masonry bed course above the top floor windows, but reinforcing the highest continuous bed courses would still provide some benefit.

Reinforcement can also be hidden behind render and plaster (see below).

Injection

Injection can be used for:

- simple crack repair
- stabilizing loose cores
- arresting delamination.

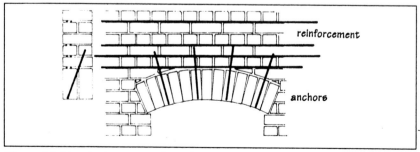

Figure 20.11   Arch converted to a beam.

For simple crack repair, the most suitable material is usually a resin-based grout. Viscosity should be chosen to suit crack size, in accordance with manufacturers' advice. Thixotropic materials can cope with a wide range of crack widths, provided there is continuous access for injection. Cracked mortar should be raked out and the gap cleaned, as for straightforward repointing. Injection is usually by gun. The face of the grout should be kept back so that the repair can be repointed. It is possible to inject cracks in rendered walls with minimal damage to the render, provided a resinous or cementitious grout of suitable viscosity is chosen. Cavity walls can also be repaired in this manner, provided the viscosity is selected to ensure that the cavity itself is not injected.

Resin injection is also suitable when the crack passes through masonry units as well as mortar. Entry points need to be drilled and cleaned prior to injection, and plugged afterwards, either by gluing a specially cut cover piece of the same stone or brick to the surface, or else by plastic repair.

Stitching and grouting cores

Solid walls with weak or deteriorated cores may be stabilized by through stitching, using rods embedded in resinous grout. Figure 20.12 shows a typical detail. The resin should flow and fill the immediate voids without dispersing so rapidly that voids are left in its wake. Horizontal and vertical spacing between stitches should be chosen on the assumption that the core has no structural value, and the stitches serve as wall ties. There are no formulae. Judgement should take into account the strength of units, measured distortions, the likelihood of further core consolidation, applied loads (a locally greater density of stitching may be appropriate under heavy point loads), and the need, if any, to share loads between the two leaves.

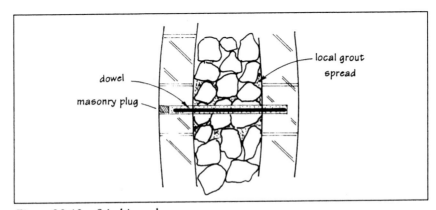

Figure 20.12 Stitching a loose core.

Figure 20.13 shows the same method, but this time the entire core is grouted. After the stitching has been installed, the core should be thoroughly washed, usually by flooding under a head of water, to remove smaller debris and to dampen surfaces ready for the grout. Cementitious grout, such as a cement:fine sand:filler suspension, is usually suitable for the permanent repair and is usually the most economic.

To ensure a good grout spread, work usually starts at the bottom. The grout is slowly injected under a light pressure, which may come from gravity feeding under low head. As the grout rises up the core, it should appear through prepared drill holes, which are bunged after the grout has established its escape. When the grout has reached all escape holes at a prearranged level, a second grout injection point (or row of points) can be opened, so that the grout spreads vertically, and the procedure repeats. This continues until the full height of the wall has been treated. The pace of treatment should take into account the bursting pressure of fresh grout (Figure 20.14). If this pressure is too high for the facing masonry, the risk of damage may be reduced by designing the stitching to take the bursting pressure as a horizontal (tie) force, ensuring that enough time is allowed for the stitches to reach their full strength before grouting starts. Alternatively, the pressure can be curtailed by the pace of grouting.

Precautions may be needed to prevent low viscosity grouts from invading areas where they are not welcome, such as service ducts and drains.

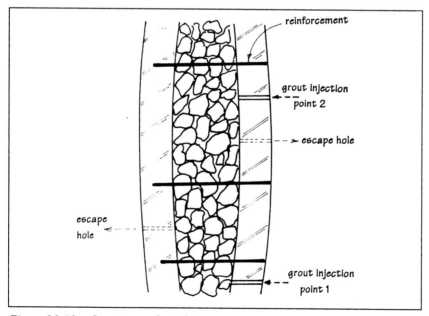

Figure 20.13   Grouting and stitching.

Stitching delaminated walls

Through stitching is often the most economic way of restoring integrity to a solid wall that is delaminating. The stitches are effectively wall ties. A stitched delaminated wall should be assessed as more slender and, therefore, less strong than the original (since it will have become two leaves) unless the delamination was only incipient and the density of stitching is high enough to recompense the slight loss of integrity.

Delamination can sometimes be spotted at reveals, if the brickwork is creeping apart from frames set into it. Without evidence of damage, however, it is difficult to detect by eye. It can be discovered by verticality measurements (Figure 20.2n, *page 201*). Both inside and outside faces should be surveyed. Delamination may be suspected if the two are not parallel. Suspicions must be confirmed by trial hole, because replastering can sometimes make an ordinary bulge seem like delamination.

When inside and outside faces have different profiles, several potential causes should be considered. The three most common groups are:
• shortening of the inside skin (rotting of bonding timber?)
• forces acting preferentially on one skin (thermal; wind; roof spread)
• delamination.

Rendering

The purpose of repair by rendering is to improve durability, strength, appearance, or all three. First, the repairer has to decide whether the alteration in appearance would be acceptable. If conservation rules apply, rendering would not normally be permitted, except for the most pressing

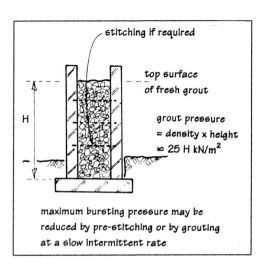

stitching if required

top surface
of fresh grout

grout pressure
= density x height
≈ 25 H kN/m^2

H

maximum bursting pressure may be
reduced by pre-stitching or by grouting
at a slow intermittent rate

Figure 20.14
Grout pressure.

reasons. There is one unusual exception to that rule. Many historic buildings now present a plain stone face because their original render has been
eroded, so a new render would restore the original appearance. Whether
the original appearance actually is wanted now is a matter for debate and
local planning authority ruling. If it is wanted, a lime render (with no
cement) or, in some cases, a lime wash would often be preferred, adding
durability but not strength.

Figure 20.15 shows a section through a typical render, emphasizing the
most important considerations. These basic requirements can be varied
to suit the background material (its porosity and smoothness) and other
qualities that would benefit the building. These include: improvement
in appearance; better resistance to rain penetration; and better thermal
insulation. If the insulation is particularly important, it may be beneficial
to sandwich insulation bats between background and render, provided
the lower strength (compared with render bonded to the background)
is acceptable.

The render can conceal reinforcement, which is potentially the most
spectacular structural benefit it can provide. This benefit should not be

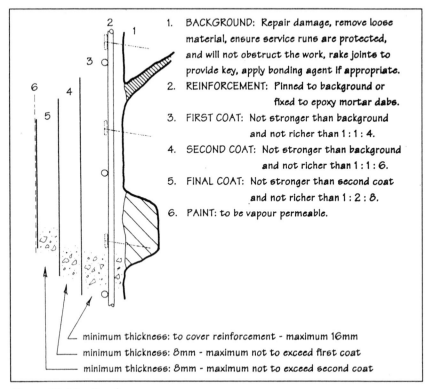

1. BACKGROUND: Repair damage, remove loose material, ensure service runs are protected, and will not obstruct the work, rake joints to provide key, apply bonding agent if appropriate.
2. REINFORCEMENT: Pinned to background or fixed to epoxy mortar dabs.
3. FIRST COAT: Not stronger than background and not richer than 1 : 1 : 4.
4. SECOND COAT: Not stronger than background and not richer than 1 : 1 : 6.
5. FINAL COAT: Not stronger than second coat and not richer than 1 : 2 : 8.
6. PAINT: to be vapour permeable.

minimum thickness: to cover reinforcement - maximum 16mm
minimum thickness: 8mm - maximum not to exceed first coat
minimum thickness: 8mm - maximum not to exceed second coat

Figure 20.15   Render.

overestimated. It cannot match the capacity of straight bar reinforcement (although the two can be combined). However, it does provide tensile capacity at all levels, making it an excellent mitigator of damage when limited ground movement is expected.

The many benefits of render make it an attractive material to introduce as part of the repair scheme. On the other hand, it does punish any lapse in standard of specification or workmanship, typically by cracking and debonding from the background – defects that are difficult to rectify. It should never be used on brickwork undergoing chemical change, for example sulfate attack or efflorescence. Another point to consider is that rendering increases maintenance costs.

Walls that are already rendered may be repaired locally to the cracks. If cracking is extensive, or other defects (such as bond loss) are contributing to widespread damage, the old render can be entirely removed and replaced by new. If the render is damaged but difficult to remove, a new render may be applied on top. In this case it will be essential to ensure that:

- the original render is still well bonded to the background, and has no other significant defects
- the rule that each coat is more permeable than the underlying one is not broken (Chapter 15, *Frost*).

Reinforced plaster

Reinforcement is a useful secret internal repair. Figure 20.16 (*overleaf*) shows four conditions that could be improved by cutting back existing plaster, fixing expanded metal reinforcement or mesh to the background, and replastering. For such repairs to be successful, the following principles should be observed:

- Cracks should be thoroughly filled first, as the reinforced plaster will not be stiff enough on its own to prevent closing movement.
- The reinforcement should be fixed to the background, using screws or epoxy mortar dabs, or otherwise kept stable until the first coat of plaster sets.
- If the expanded metal, or mesh, has a strong and a weak direction, the stronger should bridge across the cracking.
- A cement:sand coat is the ideal cover for the reinforcement, with a finishing coat to match the existing plaster as closely as possible. Where this is not practicable, the strongest mix should be used, subject to compatibility with existing materials and the need to avoid gross shrinkage. Otherwise the repair will not be able to work at normal strength.

- Services, such as electric cables, running within the thickness of the plaster, may be too much of an obstruction to make the detail practicable. In this case, an alternative such as resin injection or mortar bed reinforcement may be preferred.

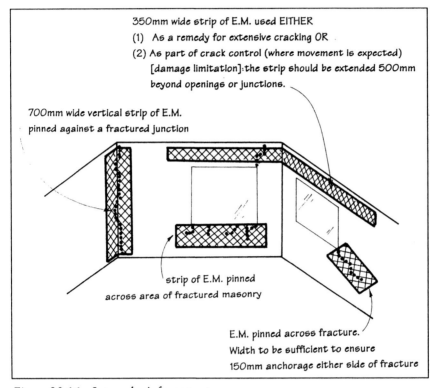

Figure 20.16   Internal reinforcement.

Chapter Twenty One

Weak materials

Weak materials are defined for this chapter as soft stone and unfired earth. When buildings were made of these materials, it was understood that regular, sympathetic maintenance would be essential to their survival. Modern owners are not always aware of this. Repairers must ensure that deterioration will not frustrate the purpose of the repair. Neither must the repair initiate deterioration by introducing incompatible materials. The other major problem to contend with is the the materials' weakness. Their small margin of safety must never be jeopardized by lax detail. This chapter discusses this problem briefly and suggests a few typical details.

Regardless of the cause of damage or purpose of repair, it is usually advisable to carry out at least a visual inspection for deterioration. Most common agents of deterioration, especially with unfired earth, are related to damp penetration at eaves, at ground level and through 'modern' render. There may also be direct wear and tear caused by vermin or vegetation.

Soft rock

Soft rock includes chalk and weakly cemented sandstone. The latter varies considerably in durability, although at the time of construction stonework was provided with adequate protection wherever it was vulnerable. Repair obviously should respect protective details and, in some cases, restore them.

Flint

Flint's strength and durability are not in doubt. It is its connection with its background that fails. The cause of failure is usually deterioration of the connecting mortar or render, particularly if it is possible for water to penetrate behind the flint and freeze. The repair detail should close this passage for water, if possible. Most flintwork consists of nodules or knapped flints set into mortar or a render face with a strong stone background carrying the wall's main structural function. Flint's impermeability ensures that hardly any water crosses the boundary between it and setting mortar, and repairs must take this into account. Bond will be weak. Repair has the best chance of success where stability is helped by horizontal support, such as stringer courses of stone or brick.

Earth walling

Strength is always a constraint on the repair specification (Chapter 15). For minor repair to a well-maintained building, it may not be necessary to establish a value for strength, provided there is caution in the detail. If the repair is major, or there has been significant deterioration, a strength assessment should be made at the outset. There are no standards to rely on, because the material was purpose made, using local knowledge. Few records were kept. Typical characteristic strengths are about $0.6N/mm^2$ (clay lump); $0.3N/mm^2$ (*in situ* clay and chalk). A modest rise in water content can slash these values. There should be a margin of safety between characteristic strength and working stresses (Chapter 4), to allow for a material factor of safety and slenderness. If strength is critical, as it might be if stresses have increased or are likely to increase, the material should be tested.

Wall thicknesses vary from 230mm for clay lump to 450–600mm for *in situ* methods such as Devon cob. (These are typical dimensions, but there are wide variations.)

One method of earth walling, known as pisé, usually achieves characteristic strengths higher than the figures quoted above. The walls were constructed by ramming the earth mix, which usually had a low or negligible clay content, between tightly held shutters, thus ensuring better than usual compaction. The compaction delivered the strength. The strength enabled pisé external walls to be as thin as clay lump. Internal pisé walls are sometimes found to be even thinner. The same problems, slenderness and strength loss, rear their heads if any restraint is lost.

Unless reliable material data are available, it should be standard practice to ensure that repair details can only widen the margin between material strength and applied stress (Chapter 4 and Figure 4.6, *page 49*).

Compatibility

Repair materials must be compatible with existing materials. Filling earth holes with brick and mortar, and replacing weak but permeable render with hard cement:sand, are repairs that are guaranteed to shorten the life of the building.

Before starting major repairs, existing materials should be analysed for particle size distribution of the soil content and for the proportion of improvers, such as chopped straw. New materials should have a similar profile, but certain minor variations can be beneficial. A modest amount of sharp sand, straw or chalk should reduce shrinkage. Strength and durability can be improved by adding up to 10% by weight of a stabilizer. If the material is of low plasticity (plasticity index below 20%, Figure 2.3, *page 8*), Ordinary Portland Cement is the best stabilizer. Material of greater plasticity may be better served by lime, or a mix of lime and cement. New rendering can improve durability, but can rarely add strength.

Although Figure 20.15 (*page 218*) shows the use of render on masonry as a strength giver, the obligation to use compatible materials rules out strong render on earth. A typical render mix would be lime:sand, or mud:sand:straw/hair. The background should be roughened to give a good key and then sprayed with water to improve the bond. If there is difficulty in achieving a satisfactory bond, it may be worth improving the prospects by using stainless steel mesh. If there is an existing render, which is found to be well bonded to the earth, it is best left alone. It can be patched where necessary, using materials similar to the existing mix.

Local experts should be able to give invaluable advice on the mix of materials.

Details

Structural stability must be assured before repair is attempted – unless, for reasons of safety or cost, running repairs must be carried out while movement is working its way out. If there is potential loss of equilibrium caused by the wall being out of plumb, it should be restrained before repairs are started. As often observed, restraint provides a healthy boost to load-bearing capacity in any case. Details of earth wall restraints are shown in Chapter 22 (Figure 22.8, *page 235*).

Narrow cracks

Cracks of up to 5mm in width can usually be healed by running in a grout of lime putty, after first dampening the crack faces and sealing the wall faces against grout leakage by masking tape or similar. A typical grout mix would be one part lime to three parts sand, with just sufficient water to ensure free flow under gravity. If the crack is continuous, one entry point will usually be sufficient. Otherwise, several entry points and bleed points can be used.

Wide cracks

The division between narrow and wide is purely practical. Grout may be suitable for cracks wider than 5mm, provided it bonds well with the parent material and does not shrink to form two cracks in place of the original one. Where grouting is not suitable, the crack should be cleaned out and dampened, ready to take an earth mix similar to the original. A little stabilizer may be added if appropriate. The mix should be well compacted, either by tamping from one face against a shutter or by working from both faces.

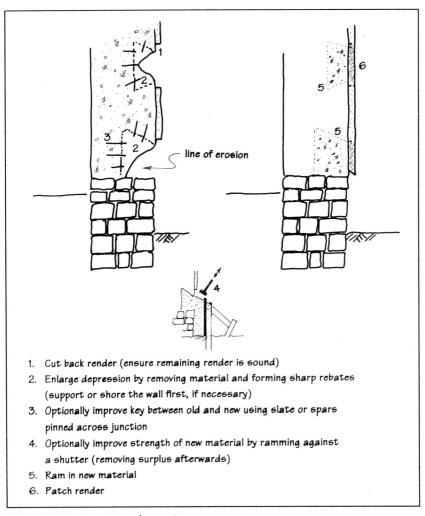

1. Cut back render (ensure remaining render is sound)
2. Enlarge depression by removing material and forming sharp rebates
   (support or shore the wall first, if necessary)
3. Optionally improve key between old and new using slate or spars
   pinned across junction
4. Optionally improve strength of new material by ramming against
   a shutter (removing surplus afterwards)
5. Ram in new material
6. Patch render

Figure 21.1   Repairing depressions.

Rat runs

Rat runs may be suspected if there appear to be small exit holes on the external face, accompanied by a fan of biological debris. The location of rat runs can be identified by thermography, which detects the small variation in temperature between solid and perforated walls. Confirmation should follow by drilling here and there into the thermographically mapped rat runs.

The runs can be grouted, but a more positive, and stronger, repair can usually be accomplished by opening the runs up from one face, removing debris, and treating the exposed hole as a *wide crack*. For predominantly horizontal runs, it is essential to do this in short lengths. Otherwise the wall may be destabilized. For predominantly vertical runs, it may be advisable to do the ramming in lifts, with pauses for shrinkage.

Depressions

Surface depressions develop as material is worn away by weather and other agents. They are healed by cutting them into a manageable shape and filling them with material similar to the parent earth. Typical details are shown on Figure 21.1.

Through holes

Through holes are treated in much the same way as depressions. Compaction, and therefore strength, can often be helped by fixing a shutter opposite the working face to provide a firm surface for ramming against (Figure 21.2, *overleaf*). In some cases, particularly when a large area of damaged or unsuitable material needs replacing, quality control can be improved by making blocks of new material similar in constituency to existing material, and inserting them into the wall, which has been prepared by cutting each defect into a regular shape with all edges sound.

If the cutting out exceeds one metre in length, or part of a heavily loaded wall (say, a pier between openings) is removed, it will be necessary to support the wall temporarily on needles.

When structural damage is accompanied by widespread deterioration, the repairer may be compelled to replace a large area of weakened material, deliberately forming a depression or through hole.

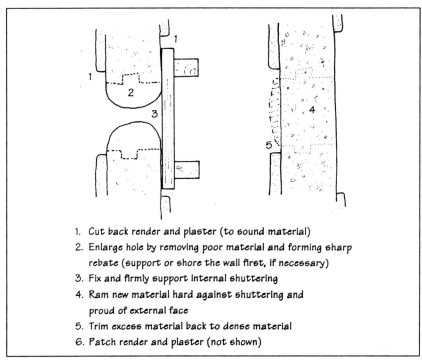

1. Cut back render and plaster (to sound material)
2. Enlarge hole by removing poor material and forming sharp
   rebate (support or shore the wall first, if necessary)
3. Fix and firmly support internal shuttering
4. Ram new material hard against shuttering and
   proud of external face
5. Trim excess material back to dense material
6. Patch render and plaster (not shown)

Figure 21.2   Repairing holes.

Chapter Twenty Two

Restraints

*To a greater or lesser extent, the roof, floors and walls of a building pro-
vide mutual restraint where they meet. Modern buildings have this
restraint by design, in compliance with building regulations. Older buildings were not always provided with planned restraint, and in many of
them the efficacy of any informal restraint has now declined or suffered
damage. The result of lost or damaged restraint is an increase in slenderness and a reduction in both the strength of individual members and
the robustness of the building. This chapter discusses means of restoring or introducing restraint where it is needed.*

Restraint and robustness

Restraint reduces the slenderness of walls, thereby increasing their
strength and resistance to buckling. It also provides a route for secondary
forces to pass efficiently through the building. These benefits combine
to produce robustness. Robustness is routinely assessed by surveyors,
who take into account:

- the building's layout (Figure 2.13, *page 18*)
- the properties of its materials (particularly strength and durability)
- its scale (storey heights, wall thicknesses, joist sizes)
- its ability to distribute forces through bearings and connections.

These factors are judged so far as is practicable without pulling the building apart. The assessment is pragmatic. There is no numerical definition
of robustness. A robust building acts as a single unit, transferring the
expected forces through itself and into the ground with negligible distortion and damage, and resisting unexpected forces (such as explosions,
impacts and gross overloading) without disproportionate harm. The
building regulations have, since 1970, required buildings of five storeys
or more to be designed to restrict to a small area any severe damage or collapse arising from irresistible force.

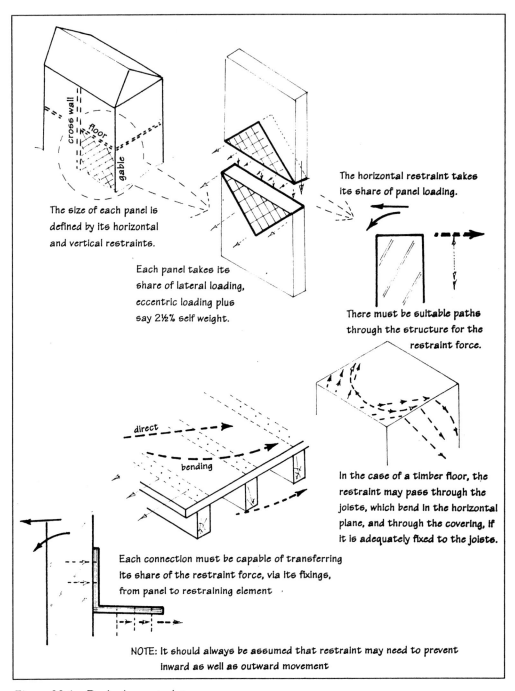

The size of each panel is
defined by its horizontal
and vertical restraints.

Each panel takes its
share of lateral loading,
eccentric loading plus
say 2½% self weight.

The horizontal restraint takes
its share of panel loading.

There must be suitable paths
through the structure for the
restraint force.

In the case of a timber floor, the
restraint may pass through the
joists, which bend in the horizontal
plane, and through the covering, if
it is adequately fixed to the joists.

Each connection must be capable of transferring
its share of the restraint force, via its fixings,
from panel to restraining element

NOTE: It should always be assumed that restraint may need to prevent
inward as well as outward movement

Figure 22.1   Designing restraint.

The easiest way to improve robustness is to provide restraint, or to improve existing restraint, where walls meet floors, roofs and other walls.

Needs of restraint

When designing new restraint, the repairer needs to establish:

- What needs to be restrained?
- What forces will be needed to assure restraint?
- What routes through the building are available to these forces?
- What methods are suitable?

Figure 22.1 follows the line of enquiry.

Practical details

The potential for damage during fixing can sometimes restrict choice. For example, bricks in poor-quality mortar may be loosened by drilling. Brittle units may shatter. Some units are too soft to take the intended fixing load (although such a problem can often be overcome by curtailing the value for individual connections, and using more of them). Hollow units will accept only certain types of fixing. Chases and notches, cut to house fixings, inevitably weaken the members, and this can be unacceptable in any material that is close to its limit of strength or stiffness.

All parts of the restraint should last the life of the building. Exposed metal, including anything within a wall cavity, should be stainless steel or other material of similar durability. In other locations, durability should at least be the equivalent of mild steel with a minimum mass of galvanizing of 940 g/m^2.

Design should acknowledge task difficulties, especially any health and safety implications. Access is often the biggest problem. Anything fixed from outside will need a safe working platform. Inside, the work can be done from an existing floor, but it will usually be necessary to lift floorboards or cut into plaster or boarding. Fixings should not clash with service runs. When choices are being made, details that are likely to be easily confounded by unexpected discovery – such as hidden damage, rot and variations in dimensions or material – should be discarded if possible.

1. MECHANICAL : DISPLACEMENT

As the fixing is screwed, hammered or fired into the materials being fixed, the latter are deformed and become locally more compact. The strength of the connection is limited by (i) the strength of the fixing and (ii) the grip of the compacted material.

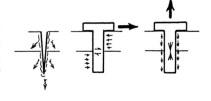

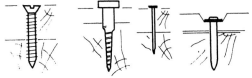

Shear and tensile capacity of individual fixings are low to moderate (with the exception of nails and pins, which should not be used in tension). Fixings may be used in groups, provided spacing is not so close that areas of compaction overlap.

Little skill is required but capacity is reduced by carelessness e.g. bent fixing, split timber

2. MECHANICAL : REPLACEMENT

A through hole is drilled to house the fixing. The shear strength of the connection depends on the strength of the fixing and the bearing strength of the material surrounding the hole. When used in shear, slight deformation may occur as the bolt shank deforms the surrounding material.

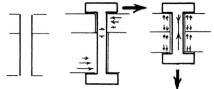

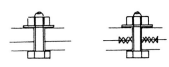

The moderate shear capacity can be enhanced, and deformation decreased, by the use of shear connectors (near left), which bite into the timber as the bolt is tightened. In tension, capacity usually depends only on the strength of the bolt.

Installation is simple but capacity can be reduced by inaccurate drilling or inadequate tightening.

3. MECHANICAL : EXPANDING

A hole is drilled for the fixing, which grips the surrounding material as the bolt is tightened against it.

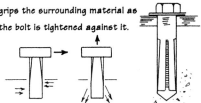

4. CHEMICAL : ANCHORAGE

A hole is drilled to take the fixing. As the fixing is installed, chemicals are released. These react to produce adhesive, which creates a bond between bolt and materials.

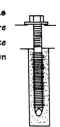

Expanding and chemical fixings are only suitable for anchoring into masonry or concrete. Both require skill in fixing. Their high capacities cannot be fully realised in weak or hollow material (or, in the case of chemical anchors, very porous material).

5. ADHESIVES

The two surfaces are glued together.

Shear capacity is high, but the connection is not usually suitable for tension.

In situ installation requires skill. It may be hazardous.

Standard text books give load capacities for nails, screws and bolts. Other types of fixing come in many varieties. Manufacturers will provide information on capacities and advice on use. Most connections are limited in their capacities by the strength of the materials they connect, and may be further limited if they are close to edges, as the material may split under stress.

Figure 22.2   Types of restraint.

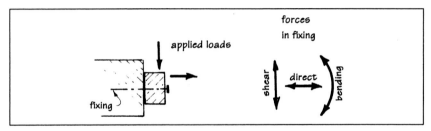

Figure 22.3   Forces in fixings.

Fixings

Figure 22.2 shows the basic types of fixings. The classification used is arbitrary.

Orthodox text books and manufacturers' literature supply all the necessary information on the capacity of fixings, but it is up to the repairer to ensure that they are used properly. Most are intended to take a shear force or a direct force, sometimes both; rarely do they expect to withstand large bending forces. Figure 22.3 shows that any load not passing through the interface between the materials being connected will cause bending. A very small bending force may therefore have to be accepted in most practical details, but it must not be allowed to get out of hand (Figure 22.4a). A different detail should be considered if that is a possibility (Figure 22.4b, for example).

It is bad practice to rely on a single fixing, because this has a small but significant chance of failure. Practicalities often dictate the use of single bolt connections, but the risk of failure should always be reduced by using several (at least three, typically at 1–2 metre spacings) to transfer the total restraint force. Then, even if one fails, the group capacity should remain adequate for purpose.

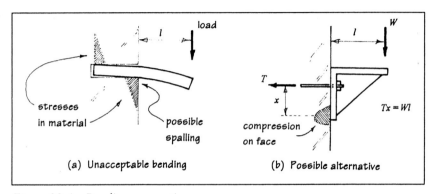

Figure 22.4   Bending connections.

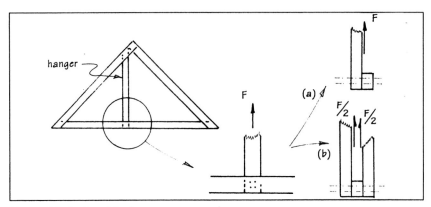

Figure 22.5   Double shear.

The capacity of through bolts is improved if they are used in double shear. Figure 22.5 shows a hanger installed to reduce the span of a tie beam. A single shear connection is shown in (a). By using a double hanger (b), each bolt transfers its load across two shear planes instead of one, doubling its capacity and, at the same time, reducing minor twisting of the timber and possible bending of the bolts.

Destination

Every force has to go to ground. As shown on Figure 22.6a, once restraint has safely passed through the fixings and into the restraining element, it must be able to find an uninterrupted route through the building and eventually downwards to ground level. In the case illustrated, the joists will be able to transfer some load to their supports by bending in the horizontal plane. Part of the load – probably most of it – may be attracted to the floor covering, because floor coverings act as stiff diaphragms provided they are adequately fixed to their supporting joists. Any interruption to straightforward bending or diaphragm action would lengthen the load path and reduce efficiency. A stairwell, for example, can make restraint locally ineffective, unless a frame is installed to guide forces around it.

The final stage of the restraint forces' journey is downwards through the walls, frames or columns that meet the floor joists and covering. In robust traditional buildings, the stresses created in these elements are very low, and need no formal accommodation. Buildings without natural bracing (sometimes because it has been removed) have to transfer lateral restraint forces over greater distances than may be comfortable. These buildings are likely to bend under normal loading and crack under exceptional loading. In tall buildings, the restraint forces created in shear walls and strong points, such as lift shafts, are a significant proportion of total (worst combination) forces, and they certainly need to be calculated.

Figure 22.6a also provides a reminder that restraint forces act in either direction, although not necessarily with equal energy. The detail must allow for this.

Tie bars, of the type that run right through the building (Figure 22.6b), can be inefficient if they rely on anchor walls to resist lateral loading unaided, thereby having to bend or rock out of plane. The tie heads can also cause damage if the transmitted force creates high local shear stresses in the masonry. Finally, a very long rod can, as it expands and contracts (typically by 1mm per 10°C change in a 10m long rod), superimpose a fluctuating force on the intended tie force. This is rarely a serious problem, but it can cause local damage to masonry in weak mortar, or any masonry that is already damaged or weakened.

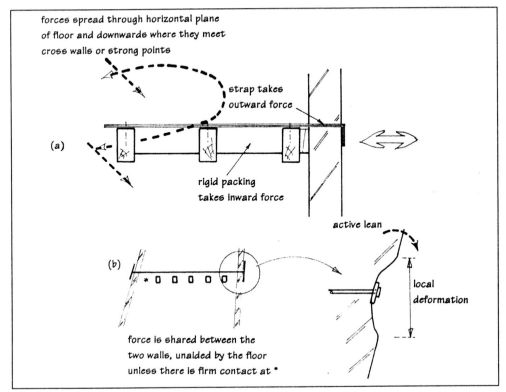

Figure 22.6   Means of restraint.

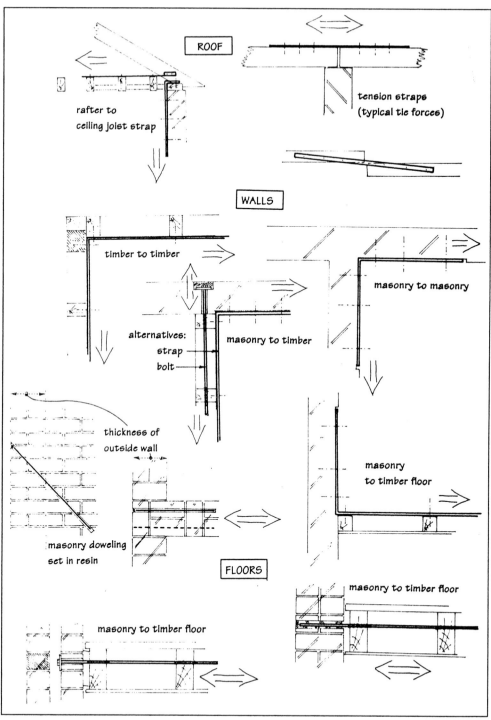

Figure 22.7   Typical restraint details.

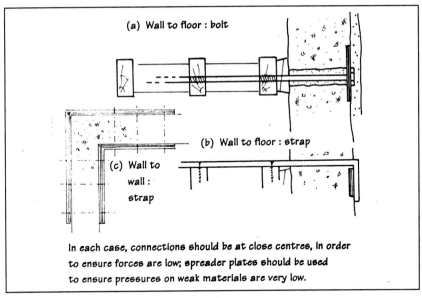

In each case, connections should be at close centres, in order
to ensure forces are low; spreader plates should be used
to ensure pressures on weak materials are very low.

Figure 22.8   Restraining weak material.

Typical details

Figure 22.7 shows some straightforward details for connecting the main
components of traditional buildings. Each would have to be adapted to
cope with specific technical requirements and task difficulties. Weak
materials, such as earth walls, require more delicate connections, of the
type shown in Figure 22.8.

PART FOUR

Management

Chapter Twenty Three

Preventive maintenance

This chapter discusses how the more costly and foreseeable defects listed in Chapter 8 can be anticipated and prevented. Savings made by preventive maintenance include not only the money that would have been spent on repair contracts but indirect costs and penalties such as permanent scars and distortions, loss of business, inconvenience to owners and users, and possibly increased insurance premiums.

Preventive maintenance and maintenance on discovery are compared in Table 23.1.

Table 23.1 Comparison between preventive maintenance and maintenance on discovery

Preventive	Discovery
Needs running costs to set up and continue.	No running costs.
Work can be budgeted.	No budget.
Client is more aware and more able to control maintenance.	Client can only react to each reported incident.
Fewer crises to manage.	More crises to manage.
Costing under control, although some unnecessary work may be included.	Little unnecessary work is done, but control over costs is poor.
Work can be planned to suit key/skilled operatives.	Little opportunity to plan.
Legal requirements can be included in planning.	Legal requirements more likely to be forgotten.
Conservation requirements can be included in planning.	Conservation requirements will need to be reviewed whenever unexpected work has to be specified.
Records of work should be useful.	Records likely to be piecemeal.

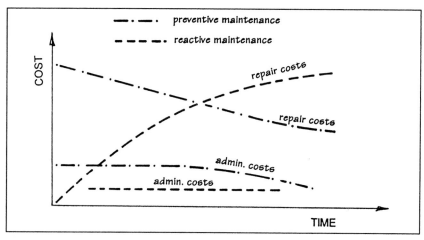

Figure 23.1   Costs of maintenance.

Preventive maintenance is justified if its long-term cost shows a saving, but this is not easy to calculate. There will be an additional administrative cost (Figure 23.1). The initial cost of the maintenance work itself will also be higher, because of action taken over and above normal maintenance, to remove faults and alter features that might otherwise mature into damage. As time goes by, the administration of preventive maintenance becomes more routine, and the cost of the work falls as the benefit of prevention is harvested.

The professional carrying out preventive maintenance needs the diagnostic skills of the repairer, and the same pool of knowledge of structural behaviour. It is useful, as well, to have a file of site-specific knowledge. This might include:

- records (drawings, reports, health and safety files, archives)
- local knowledge (endemic problems such as flooding or slope instability)
- soil data (preferably borehole data but drift geology and soil survey maps are also helpful)
- site layout (positions of service runs and entries, paving, vegetation)
- regular reports (inspections of vulnerable parts, damage investigations).

Risk assessments

Employers in charge of commercial and industrial premises have a duty to assess the risks to anyone affected by the work activities. They must:

- identify hazards
- identify who is likely to be harmed, and how
- decide on precautions
- record and periodically review the assessment.

Such risk assessments may share some of the tasks of preventive maintenance, by requiring, for example, periodic inspections of vital structural
elements, strengthening of emergency routes, protection of parts vulnerable to vehicle impact, and so on.

Hazardous processes

Some processes incorporate direct hazards to structures, and require rigorous assessment. For example, dust explosions are a generic hazard in certain industrial processes involving food, chemical and wood products,
among others. If the dust is combustible and can float in the atmosphere in
sufficient density, and ignition takes place, there is an immediate flaming,
which has the same effect on the structure as an internal blast. There can
be a secondary effect, if the flame can travel to other parts of the building and can find dust (for instance on tops of cupboards and ledges), which
will act as fuel, encouraging the explosion to spread and develop into a fire.
   Typical options for reducing the risk of explosion and fire are:

- confining the dust to the process rooms if possible
- regularly removing dust from other areas if 100% confinement is not
  possible
- reducing sources of ignition
- providing venting for the potential explosion adjacent to its source.

Processes change and precautions can lapse, so regular reassessment is
essential.

Typical causes of damage

Preventive maintenance can also include enquiry into the typical causes
of structural damage (Chapter 8). One example is subsidence resulting
from erosion by water. This is discussed in Chapter 9, and illustrated in
Figure 9.2 (page 82). The conditions essential to the operation of this
cause are, of course, erodable soil and a source of water. If the foundations
are on silt, fine sand, low-grade chalk or made ground – or if there is a likelihood of swallow holes in the area – then there is a potential for damage
caused by escape of water. This would prompt an enquiry into potential

Table 23.2   Types of drain material

Pipe/joint material	When most in use	Properties relevant to life span
Clay, salt glazed C:S joints	until mid 1960s	Easily fractured by minor ground movement
Ductile iron with flexible joints	from 1950	Pipe lengths are long and may fracture in unstable ground. Rodent proof. Can be supported between supports.
Pitch fibre with taper collar joints	1960–70	Pipes easily deformed or punctured by poor bedding. Joints easily sprung by pipe movement, causing leakage; therefore susceptible to moderate ground movement.
Clay, salt glazed rubber joints	from 1970	Flexible unless bedded and haunched on concrete, in which case easily fractured by minor ground movement.
Vitrified clay pipes with plastic sleeve and rubber gasket joints	from 1980	Joints are flexible but long individual pipes are brittle and can be fractured by moderate ground movement
Pitch fibre with plastic joints	1975–90	Flexible. Tolerant of ground movement but easily deformed or punctured.
Plastic pipes flexible joints	from 1975	Flexible. Provided the run is evenly bedded, damage and leakage is likely to be caused only by severe ground movement or accident.

sources of water close to, or within, the building, including storm water drainage and soakaways, water mains, and foul drains. With foul drains, for example, the most hazardous spots are where the drains exit the building. The likelihood of leakage is greater in certain types of drain material than in others (Table 23.2). A closed-circuit television survey, with a smoke or water test, would establish the condition of the drains, including any current leakage. When the potential for failure of the foul drains and other water sources has been assessed, preventive maintenance can be planned, if the expenditure is justified by the risk.

Chapter 9 describes the ground conditions most likely to cause structural damage. It is not difficult for an experienced professional to decide which conditions might apply to the building in question and to carry out assessments where appropriate. Defects in the superstructure tend to be typical of the period. Medieval timber-framed buildings, for example, may have decayed sole plates and other timber decay wherever damp has penetrated, as well as distortions caused by alterations or overloading the floors. It should be possible to compile a list of potential above-ground hazards to go with those below ground, and to assess the risk of damage where the cost of doing so is worthwhile.

Health and safety

A good deal of maintenance work is too small to require either the appointment of a planning supervisor under the Construction (Design and Management) Regulations, or the preparation of a health and safety plan. Nevertheless, requirements on designers always remain in force (Chapter 6). Table 23.3 (*overleaf*) lists just a few of the problems that are common to maintenance work. Designers of maintenance work are often more familiar with the environment than those who will be carrying out the work. This provides them with all the more opportunity to make the place of work as safe as is reasonably practicable. Consider again the example of a flat roof congested with service runs and small plant rooms (Figure 23.2). All work – inspection, cleaning, maintenance, alteration – would have to include precautions against falls, and these precautions would add to the cost of the work. As soon as funds can be justified, a surer and more economic plan would be to provide permanent safety barriers, safe walkways and safe lights, which would satisfy most future health and safety requirements for roof work. The cost of repeated temporary work would be avoided.

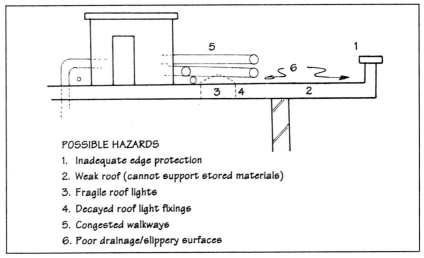

POSSIBLE HAZARDS
1. Inadequate edge protection
2. Weak roof (cannot support stored materials)
3. Fragile roof lights
4. Decayed roof light fixings
5. Congested walkways
6. Poor drainage/slippery surfaces

Figure 23.2   Roof working hazards.

Table 23.3   Maintenance hazards

Hazard	Possible precautions
Roof working *(falls)*	Provide edge barriers. Replace or cover fragile lights/openings. Provide safe walkways. Provide crawler ladders for work on slopes. Provide safety harnesses or fall arrests if a risk remains.
Other working at heights *(falls)*	Provide a tower or similar safe platform in preference to a ladder where possible. Provide barriers or safe covers at all edges (e.g. stair wells, shafts).
Noise *(ear damage)*	Reduce noisy tasks. Provide silencers. Provide ear defenders. Cordon off noisy activities.
Dust and fumes *(respiratory disease)*	Regard all dusts and fumes as hazardous, especially in confined spaces. Check whether the likely dust is especially hazardous (HSE Publication EH40), and if so take special precautions. Reduce tasks creating dust or fumes. Use equipment that extracts dust or exhaust fumes at source. Plan tasks (e.g. damping surfaces, providing ventilation). Provide protective equipment including respiratory equipment when necessary.
Asbestos and lead *(severe disease and poisoning)*	If there is any possibility of disturbance, carry out a survey and plan the work in accordance with regulations (Chapter 6).
Striking services *(injury and electrocution)*	Locate and record before starting work. If necessary, work when the building would otherwise be empty and isolate the service.
Congestion *(injury)* (other trades and other users of the building needing the same area)	Reduce by planning. If necessary, work when the building would otherwise be empty. Alternatively, issue a permit to work that controls the timing and imposes conditions that reduce hazards.Inform all people affected. Consider how the work might interfere with existing emergency procedure. Plan to fit in with existing procedures.

Note: Hazards are unique to each job and precautions should be tailored to each job.

Chapter Twenty Four

Planning and conducting work

This chapter discusses the investigative work required for high levels of intervention, and some of the problems that arise during refurbishment.

Every building, no matter how enduring it may appear, sooner or later will need some form of intervention. Table 24.1 shows a hierarchy of intervention. Divisions and definitions are not watertight.

High levels of intervention, such as refurbishment, are driven by economics. The value of the refurbished building is expected to be higher than its current value plus cost of refurbishment.

Table 24.1   Hierarchy of intervention

Refurbishment	Extensive renewal, which may be a mixture of part demolition, repair, strengthening and restructuring, usually to renew the economic life of the building or to meet higher standards.
Renovation	Piecemeal renewal, which may include part demolition, repair, strengthening or restructuring, usually to improve standards in part of the building.
Repair	Replacement or strengthening of materials damaged by decay, overloading, movement or other cause.
Maintenance	Inspection, cleaning, minor replacement and other attention intended to keep the building in use and able to provide the required level of service for a specified or indefinite period.
Preservation	Sufficient attention to keep the building safe and free from rapid decay, with no effort to keep it economically useful.
Controlled deterioration	Limited protection against collapse, rapid decay or vandalism, either because the building has no more use or as a temporary measure while its future is considered.
Demolition	Final removal, leaving the site available for a new building or other use.

Reasons for refurbishment

The most common reasons for refurbishment are as follows:

- *Dereliction:* the building cannot be brought back into use by simple repair or maintenance.
- *Calamity:* severe damage, such as fire or accident, has made the building unfit for its purpose
- *Obsolescence:* the building has become unfit in one way or another – in its thermal or acoustic performance, the efficiency and range of services, or compliance with legal requirements – and it can only be brought up to date by substantial alteration.
- *Comfort:* the quality of accommodation may have fallen behind current standards.
- *Change of use:* the building's current use is no longer economic and it needs to be converted to something more useful.

There is often a mix of reasons. For example, an old mill may have shut down some time ago with no possible future as a mill. Before a new use is found, it becomes derelict and, in turn, is preserved, repaired and refurbished on its way to becoming a concert hall.

Types of refurbishment

The structural content of refurbishment may not arise directly from structural failure. It may be a response to new demands: if a lift is installed, it may need a pit; new fire regulations may require an additional staircase plus support and framing; essential new equipment may need enhanced support. Alternatively, the structural content may be a response to space requirements: partitions need to be removed; a mezzanine floor is proposed; a generous roof space wants converting to office use.

The most extreme form of refurbishment, just short of total demolition, is the retention of the facade only, with a complete rebuild behind.

Structural advice

The building professional may be asked to provide, assess or interpret structural advice at any of the following stages:

- urgent appraisal
- urgent repair
- investigation
- monitoring
- feasibility studies
- firm proposals
- project planning
- work in progress.

Urgent appraisal

An immediate judgement has to be made if the owner (or organization with statutory duty) perceives a serious imminent danger. This can be trusted only to someone who is both qualified by experience and confident in the role. The occasion may be a singular occurrence, such as a fire or impact, or it may be the discovery of perilous decay. On arrival, first thoughts are for safety. If there is danger, the public should be denied access by barriers. Authorized visitors (who may need to inspect or record parts of the building) should be warned away from dangerous areas by barriers and notices. Having considered risks to the public and to authorized visitors, the investigator should assess the risk of personal injury during the appraisal itself, and should take any necessary precaution.

Urgent repair

Anyone expected to provide an urgent appraisal should be allowed freedom to take immediate action to assure the short-term structural safety of the building. It is usually convenient to specify temporary weatherproofing and security at the same time. Typical measures include:

- propping damaged or weakened areas – or areas temporarily overloaded by debris
- shoring walls whose stability is in question
- providing a temporary roof
- blocking openings that are no longer weatherproof
- securing the building against entry
- deterring vandalism.

'Temporary' measures can remain in place for a long time, often much longer than foreseen when the urgent appraisal is made. Falsework materials are expensive to hire, so it is worth considering buying the necessary equipment and selling it on later. Connections between temporary works and the building should not be of an *ad hoc* quality, liable to deteriorate quickly. Weatherproofing material should be firmly supported.

Investigation

Provided the building is safe (or has been made safe by urgent repair) the investigation can be carried out without undue pressure of time. It is not too early to involve the local planning authority, especially if the building is historic. In this case, the principles of conservation will apply to any opening up or testing. Furthermore, the local planning authority's policy

towards refurbishment may influence the investigation itself. For work
of high intervention, especially if the building is historic, it is worth planning a wide range of tasks, encompassing most of the following:

- records search
- visual inspection and distortion survey
- non-destructive inspection
- non-destructive testing
- opening up
- destructive sampling and testing
- recording findings.

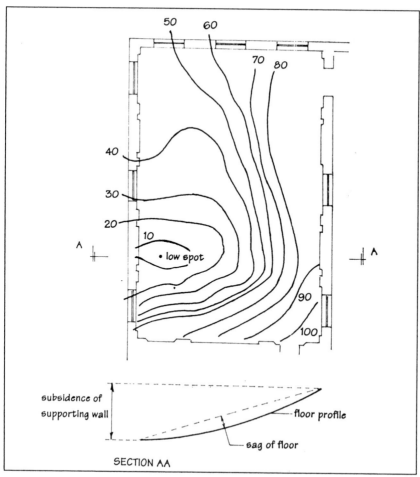

Figure 24.1   Floor distortion survey.

RECORDS SEARCH

It is worth considering Building Control's archives for recent buildings or recent work; county archives for important historic buildings; owners' files; and even standard details of similar construction that may be lodged in institution and specialist libraries.

VISUAL INSPECTION AND DISTORTION SURVEY

Visual inspection should achieve as much as possible before using any other techniques. Those skilled in diagnosis will achieve the most.

Distortion surveys are particularly useful. They can reveal unexpected sources and surprising degrees of movement. Figure 24.1 shows contours on a 10m-wide timber floor. The contours identify a local, but substantial, movement of the floor support. The strength and precision of the evidence justified opening a trial hole through granite sets in a narrow, busy, public access. A trial hole situated at a less publicly inconvenient spot would have provided much less useful information.

Distortion can sometimes date movement. For example, if the masonry beds of external walls are off level but a replaced window remains square, that would suggest that movement of the wall predated the work to the window. An off-level re-levelled floor would, on the other hand, usefully advertise a persistent problem.

NON-DESTRUCTIVE INSPECTION

Various aids to visual inspection do not harm the structure or its coverings. Technology is developing rapidly, and when a need is perceived it is worth speaking to specialist suppliers of equipment and services. Examples are:

- *CCTV cameras* for inspecting pipes, ducts and tunnels
- *pulse radar* for establishing thickness of materials and detecting voids
- *thermography* for detecting delamination (such as render debonding), tunnels and hollows
- *resistivity* for locating pipes and cables
- *borascopes and endoscopes* for inspecting cavities (not strictly speaking non-destructive, but the necessary entry holes can usually be small, discrete and easily repaired)
- *metal detectors and ground penetrating radar* for detecting underground metal, plastic, clay and concrete pipes.

NON-DESTRUCTIVE TESTING

Load testing can usefully supplement modelling and analysis. It is time consuming, however, even for the simplest test. Anything complicated requires specialist advice. The most important point, with both simple and complex load testing, is to ensure that no unnecessary or forbidden permanent damage is caused. The degree of permitted distortion should be predetermined, and behaviour should be under continuous review throughout the loading.

OPENING UP

In many buildings, virtually all the structure below eaves level is hidden behind render, plaster, floorboards and ceiling boards. Although coverings register damage and distortion, nothing compares with looking at the real structure. If loading is to be increased, it will be essential to see enough of the structure to model its performance. That means cutting out, which is expensive, inconvenient and sometimes difficult to reconcile with the principles of conservation or the nervousness of the owner. Viewpoints should be chosen carefully, not at random. The investigator should judge where problems are most likely to be found. With a wall, this might be behind damaged plaster, at the point of maximum distortion, or at junctions that may have debonded. With a timber floor, it might be at points of damp penetration, likely notching or at mid-span (maximum bending stress). It very much depends on what clues are already available. It also depends on what compromises have to be made to reduce inconvenience and damage. In the case of historic buildings, the views of the local planning authority should always be sought before any cutting or probing is done.

OPENING UP FOUNDATIONS

If the proposed refurbishment will affect foundation loading by more than a few percent, it will be necessary to open trial holes. The same considerations apply as for opening up the superstructure; the same compromise has to be made between choosing the most critical areas and causing the least damage and inconvenience – and cost. Most trial holes are opened externally and we can be less inhibited about them than we can about internal holes, even if the building is historic. If archaeological evidence might be disturbed, the local planning authority should be informed.

Trial holes should expose the foundations at formation level. The interface between foundation and subsoil is always of particular interest. A gap or uncharacteristic looseness of the ground here will usually be a sign of foundation or subsoil movement. It is usual practice to auger a bore-hole, starting at foundation soffit level, and logging, testing and sampling the soil to a depth beyond its likely influence on building behaviour.

SAMPLING AND TESTING

Sampling and testing provide information on material strength. In build-ings constructed earlier than the mid-twentieth century, there is likely to be a considerable scatter of results. The older the building, the more the scatter, because materials were not usually ordered to strength, and quality control in their manufacture was not focused on strength.

Sampling does enable other information to be gathered, most usefully on the detail and condition of structural members.

RECORDING FINDINGS

Clearly, records of observation should be impartial and comprehensive. This is more difficult than it sounds, because every task has a particular purpose, and it is natural, under pressure of time, to restrict observation to that purpose. On many occasions, however, there will be other features that are of little interest on the day, but may become important later. If timber is exposed for the purpose of inspecting rot, the presence of other defects (worm holes, notches and so on) and anything else incidentally revealed (joints, fixings, fittings) can be measured, photographed or described.

Recording adds to the building's log book (or perhaps inaugurates it). At a later date, the information may serve a new purpose. It may, for example, help establish whether deterioration has been slow or rapid.

All investigation of historic buildings should be preserved in a report, which can be included in the log book, with a second copy archived else-where. Anything that is removed as part of the investigation should be photographed, even if the intention is to replace it. If the proposed refur-bishment is expected to alter the appearance or structure in any way, drawings and photographs of the current details should be made.

Monitoring

Monitoring is often used to find the cause of damage or to discover whether it is still active, but results take time to accrue. If this delays progress on design and construction, it can be an expensive luxury. In many cases, it would be better to put sufficient effort into the investigation to enable design and planning to proceed with confidence and without delay.

If monitoring is considered worthwhile, it is essential to be clear about its purpose. Simply confirming that movement has taken place is of little use. Results must unambiguously demonstrate what has happened: the scale of the movement, its direction and what was causing it. They must be sound enough to promote a decision on structural repair.

Buildings that are fire damaged, or otherwise in a poor condition, lie empty during the period between investigation and start of refurbishment. This provides a convenient 'float' when monitoring can be carried out without delaying anything. It is a hazardous period for the building. Further deterioration can take place, especially through rot and theft, leading to further cost and in some cases irretrievable loss and damage. These risks can be monitored. Although such monitoring consists of informal inspection, it should be planned so that significant problems are recognized as early as possible. Each visit should begin with a general inspection and should then follow a checklist, which may be expanded from time to time. The checklist might typically include inspecting:

- damp and poorly ventilated areas, where rot may take hold
- all temporary supports where minor movement may loosen wedges or connections
- foundations of temporary supports
- temporary weatherproofing.

Security is important. Unauthorized entry can damage the building and the trespasser. Where practicable, entry should be prevented by a perimeter fence, as for new construction work. The level of security should be proportional to the importance of the building and the potential for damage and theft. It should deter vandals by looking impregnable and durable. All accessible windows should be boarded with strong, preferably new, material. Glass should not be left broken and, if necessary, should be protected by grills.

Feasibility

The mix of organized and apparently incidental information obtained from the investigation and any monitoring should enable the designer to produce and analyse an appropriate model of the existing structure, and credible models for future options. The worth of each option will depend on its structural efficiency and buildability, health and safety issues, and the many non-structural constraints including funds, fire regulations, service requirements and performance (thermal, acoustic and weathertightness). The feasibility study ends when the most attractive possibilities have been developed to the point where their costs and benefits can be compared.

Firm proposals and permissions

In some cases, no scheme justifies its outlay. Assuming that at least one scheme is justifiable, the designer can embark upon the preparation of the contract documents and working drawings. The choice of contract conditions and the style of the specification will be influenced by:

- whether the building will remain occupied during the work
- whether space has to be shared with neighbours during the work
- to what extent amendments are likely to be generated by what is discovered as work progresses
- the quality of craftsmanship available
- the views of the local planning authority.

The last is not intended to be the least. As mentioned more than once, the views of the local planning authority should be sought at the earliest opportunity and taken into account during the investigation and feasibility study; otherwise a lot of detailed preparation can end up as waste paper.

Planning the work

If the building is in daily use, a choice has to be made between finding temporary accommodation for occupants and planning the work around them. There is a natural reluctance for businesses to re-accommodate if work is humanly possible during occupancy, but sometimes a high price has to be paid for staying put. Both staying and going have cost implications that are difficult to calculate, and easy to underestimate, until some realistic planning has been done.

With any project, one of the early acts of planning is to establish site rules for daily procedures and emergencies. These must be merged with the users' rules if the building is to remain occupied. Any adaptation of users' rules must be communicated to all occupants.

Refurbishing a building in use demands constant communication. Users must be told in advance of any changes that need to be imposed on their routines. Construction operators must be sympathetic to users' needs and must not be as boisterous as they might be on a new construction site. The programme of work will be a compromise between the needs of contractors and the needs of other users, and will therefore seem disruptive to both. There may have to be set times for certain tasks to be carried out. Some areas, for example, might be entered by operatives only during weekends or shutdowns; hazardous and noisy work may have to be deferred until outside normal working hours; deliveries may need their own strict timetable. This sometimes means that progress can be made only during windows of opportunity. Missing one window and waiting for the next can be disproportionately detrimental to the programme – something else to communicate to users as soon as possible.

Whether or not the building is occupied, all planning has to account for the fact that work cannot follow the logical sequence of trades that new construction can. If operatives can be flexible, with sufficient talent and experience to work across trade boundaries, they will make an invaluable contribution to project efficiency.

Planning and programming must also acknowledge areas of uncertainty. Reasonable time and expertise should be allowed for handling unforeseeable problems as they arise, to avoid every problem becoming a drama.

Inspection of work

Time should be allocated in the designer's programme for inspection on demand (often when problems are discovered), for carrying out additional investigation at short notice, and for designing variations. Variations may sometimes be difficult and urgent, so channels must be kept open for discussion with decision-makers, including the client and local planning authority, and for negotiating costs, securing funds, reprogramming the work and, not least, informing everyone who might be inconvenienced.

Problems and solutions

With refurbishment, there is a greater temptation than with lesser work to introduce substantial reinforced concrete and steel components into brickwork and timber buildings. If this cannot be avoided, the difference in structural behaviour between old and new should be recognized as a design constraint.

The temporary works should be regularly inspected until they are dismantled.

Uncovering unexpected details may present opportunities as well as difficulties. This may draw the client into daily decision-making, which can be dangerous to the budget and the programme. As already noted, planners must also have the opportunity to view important discoveries.

When little but the facade is retained, inspection of what remains must be thorough, because its weight, dimensions, details of construction and, above all, any current defects or distortions, are critical to the design of the foundations and the restraint offered by new to old. The foundation design should normally aim to avoid significant differential settlement between new and old, because the latter would be likely to suffer more. This is difficult to achieve, except by assuring a lower than normal settlement for both, inevitably at some material cost. Alternatively, it may be possible to combine new and old foundations by underpinning the facade (and any other retained features) and extending the underpinning to act as a base for new columns and walls. This would not eliminate differential settlement, but it might be the most efficient way of reducing it to acceptable limits.

The connection between the facade and new construction must allow for differential movement, without losing essential restraint. This can be achieved by connections that restrain lateral movement, but allow limited freedom to move vertically and longitudinally.

Technical discussion of facade retention is beyond the scope of this book. Such a high level of intervention would normally be under the control of an experienced specialist throughout its design and construction.

Chapter Twenty Five

Insured perils

This chapter enlarges on defects that may be covered by building insur-
ance, in particular subsidence, heave and landslip. It deals with some of
the technical issues that arise during subsidence claims. Two terms are
used a little loosely. 'Water demand' will mean the amount of water the
tree (or vegetation) is able to remove from the soil at the site in question,
and 'pruning' is taken to include pollarding, crown thinning or any reduc-
tion of leaf area. As these inaccuracies imply, the advice of an arboricul-
turist will often be more than useful in choosing repair options. If these
options include tree management, the practical work can be carried out
by a tree surgeon; but professional advice is best supplied by a qualified
arboriculturist. A few other terms relevant to building insurance (such as
'settlement') are discussed in the Glossary.

Diagnosis

As previously discussed (Chapter 7), no symptom on its own – no single
piece of evidence – can provide proof of cause. Diagnosis may be confirmed only when all the bits of evidence are tested by indicators and we
can be satisfied that every potential cause has been eliminated except
the true cause or causes. Fortunately, the indicators of insured perils are
fairly distinctive. For example, Figure 25.1 shows the case where a level
survey suggests that any cause other than heave is unlikely. With additional indicators provided by the timing (onset and progress of damage),
existence of vegetation (previous and current), pattern and scale of damage, verticality profiles, and so on, there should be little doubt in the end
about what actually happened. This is the normal case. It is, of course,
impossible to eliminate the very unusual (see, for example, Figure 7.6,
page 69). But if the procedure outlined in Chapter 7 is consistently pursued, mistaken diagnosis should be unusual and very rarely disastrous.

Prognosis: clay subsidence

Figure 25.2 (*overleaf*) repeats the generic, and therefore simplistic, diagram used in Chapter 9 to show how the influence of a single tree increases and diminishes. As noted in Chapter 9, on many sites where vegetation grows close to buildings, no movement is discerned (not even stage A is reached), and few cases reach stage C. Further to this point: if the offending vegetation is removed soon after the discovery of damage,

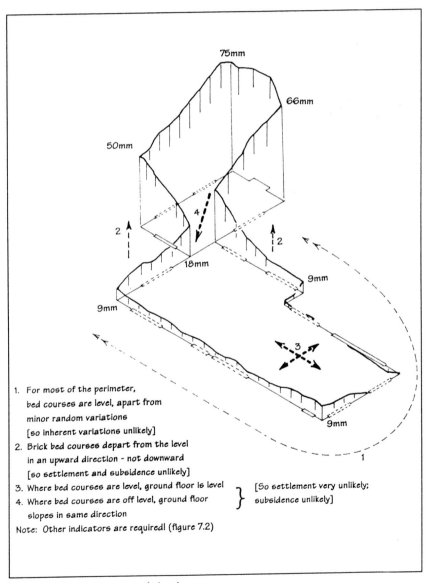

1. For most of the perimeter,
   bed courses are level, apart from
   minor random variations
   [so inherent variations unlikely]
2. Brick bed courses depart from the level
   in an upward direction - not downward
   [so settlement and subsidence unlikely]
3. Where bed courses are level, ground floor is level
4. Where bed courses are off level, ground floor
   slopes in same direction   } [So settlement very unlikely; subsidence unlikely]

Note: Other indicators are required! (figure 7.2)

Figure 25.1   Indicators of clay heave.

recovery is quick and repair costs are modest, pleasing client and funder. Recovery takes longer after the progressive stage has been reached. This is illustrated in Figure 25.3.

Repair options

Options for intervention and remedial works are summarized in Figure 25.4. The repair-only option should always be preferred except where any of the following circumstances apply:

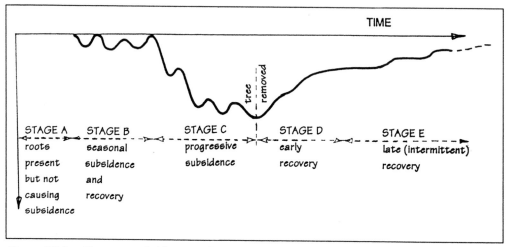

Figure 25.2   Clay–tree subsidence and recovery.

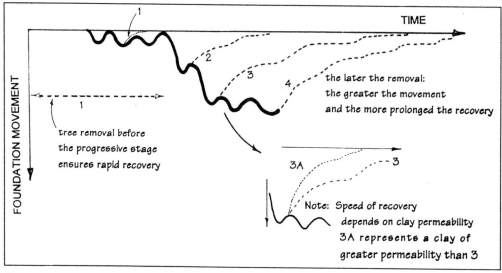

Figure 25.3   Timing of tree removal.

- Other causes also operate and demand additional attention.
- The building has become dangerous and needs urgent rescue.
- The estimated time for recovery (Figure 25.3) is unacceptably long.
- The cost of repeated repairs during a vigorous or prolonged recovery is unacceptably high.
- Vegetation removal is not permissible for legal or technical reasons.

When less than ideal circumstances prevail, the obvious alternative to repair-only is underpinning (Chapter 17). It is a drastic step. Unlike tree control, which can be monitored and modified if necessary, underpinning should be regarded as a once-and-for-all solution. The possibility of having to extend or deepen it at a later date should be avoided at all reasonable costs. (Considerations of liability may cap funding at a level below apparent best value for money from the technical point of view.)

Underpinning schemes should be designed to cope with the following circumstances:

- All parts that previously moved should be underpinned; only stable areas can be left to existing devices.(But note that every real case has to be designed on its merits, and there are exceptions to every rule. Sometimes the underpinning may be stopped short of areas where subsidence was small and the cost or inconvenience of underpinning would

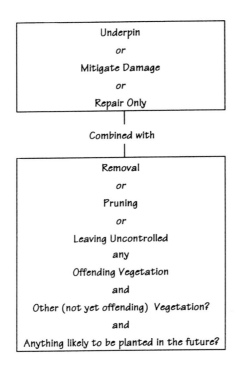

Figure 25.4
Clay subsidence options.

be large. A case in point might be a ground floor slab whose mild recovery would be less disruptive than its replacement.)
- The underpinning must contain adequate anti-heave precautions, so that the building is not damaged by subsequent soil recovery. (These precautions should be installed whether or not vegetation control is carried out, on the assumption that roots will be cut by the underpinning. In any case, the vegetation is likely to die before the building becomes obsolete.)
- The underpinning must be proof against future vegetation growth. (Only a total wall and ground floor underpinning on deep piles will guarantee stability in the face of any conceivable vegetation growth. The cost of this would not normally be justified, but the cost of anticipating further growth of *existing* vegetation might be – subject, as always, to liability issues.)

Underpinning should be deep enough to take the foundations below any existing moisture deficit. If allowance is made for future growth of vegetation, it is tempting to refer to NHBC Regulations, which determine safe depths by tree species, distance from tree to building and soil conditions. The temptation should be resisted. The logic is superficial. NHBC tables for foundation depth will keep buildings safe from all but exceptional conditions. If, however, subsidence is so bad that underpinning is needed to arrest it, then conditions are, by definition, exceptional. With reference to Figure 9.7 (*page 87*), we are in the shaded tail beyond characteristic root growth, where NHBC advice would be unsafe. Unfortunately there are no guidelines of similar status that can be referred to for exceptional conditions. Decisions can be based only on the delicate combination of personal experience and site-specific knowledge. To the extent that one tree on a site has exceeded its 'normal' limits, every other tree on the site has the potential to do the same, in its own style. The difficulty of judging events of such low predictability often tips the balance in favour of using piled underpinning in preference to other methods, because piles are less sensitive to the uncertainties of variation in soil water content.

The repair-only and underpinning options are at opposite extremes. If the limits of future movement are known, it can sometimes be cost effective to supply tensile resistance to vulnerable points (see Figures 17.4, 20.15 and 20.16, *pages 160, 218 and 220*). The reinforcement cannot stop the movement, but it can reduce the damage caused by it. Unfortunately, because the forces imposed by the soil during subsidence and recovery (and heave) are not predictable, mitigators can only be designed empirically (Chapter 9, *Building performance during recovery*). They cannot be accompanied by any guarantee, other than the assertion that the level of damage will be lower than the unknown, unverifiable level that would occur in the absence of mitigation. The option of mitigating the damage is, therefore, not often the first choice of either owners or funders; it does, however, tend to be more acceptable when the owner *is* the funder.

Pruning and water demand

Pruning immediately reduces the water demand of a tree. If it is in poor health, or the pruning is too severe, the tree may be permanently damaged. If it is in good health, more often than not it will re-establish its previous water demand within a very few years. Only careful and regular pruning, therefore, can permanently suppress water demand. An arboriculturist can advise on the best specification and schedule for pruning, but all the unpredictable factors will remain. No one can guarantee a particular outcome, however earnestly the instructions are pursued.

Pruning and subsidence

Case history shows that, once a tree is causing subsidence, pruning its leaf system can only reduce future movement, at best; it cannot eliminate it.

Root pruning would eliminate the demand for water beyond the reduced perimeter of the root system, but it is a crude operation, dominated by the need to cut off all roots found on the wrong side of a specified vertical plane. A line can be specified for the excavation, but its length and depth must be verified on the day. If it has to go deeper than expected, it could endanger the building (Figures 3.10 and 17.1, *pages 38 and 158*) or the tree. A barrier might be installed within the excavation, to obviate the need for repeated root pruning. Unfortunately, it would have to go far beyond the limits of existing root growth, because the very act of replacing a deep slice of clay by a barrier introduces conditions (oxygen supply in particular) that promote root growth. There is more than a slight risk that roots will be encouraged to spread deeper and wider than previously. On one occasion, an otherwise well-placed root barrier was stopped 250mm below ground level so that a drain could be run along its top. Roots exploited this weakness with such alacrity that it was possible to imagine them curbing their impatience as the drain layers finished their contract, tidied up and left site.

Tree control

Success (meaning zero risk of future subsidence) can be guaranteed only by felling every tree in sight. This is a policy that hardly anyone would like to see adopted anywhere. Preventive tree control will normally be a compromise.

It is possible to imagine an obvious case for felling. Most professionals would agree that a vigorous maturing tree of high water demand, growing in clay of extremely high plasticity, within a metre or two of an old building with shallow foundations, would represent a severe risk. They would also agree that, as each factor recedes from its value of maximum menace, the risk becomes lower. Unfortunately, it does not do so predictably; and there is only a brief transition from the obvious to the unreliable. In theory, it should be possible to set up formulae to give yes–no decisions on felling and pruning, but nothing seems to pass the practical test of producing the right answer 90% (or other acceptably high percentage) of the time. Perhaps something reliable will eventually emerge.

Meanwhile, preventive tree control can only be a site-by-site decision, guided by shrinkage potential and the other factors used in NHBC calculations, but strongly directed by local knowledge of history, geography, and botany:

- Have there been previous cases of subsidence in the area (if so, how many, how severe)?
- Do topographical and surface conditions encourage soil desiccation?
- Is there, for example, an impermeable hard standing that may delay rewetting and exaggerate the effects of roots?
- What can be said about the age, condition and future prospects of individual trees?

This line of enquiry provides an informed but inexact opinion. It would be legitimate to take other factors into account before making a decision, including how important it is to prevent damage. If the building is historic, sensitive or uninsured, or its daily use would be seriously disrupted by subsidence and recovery, it should be offered a high level of protection. Less compelling cases might be trusted to face a little more uncertainty. Where it is practicable and acceptable to do so, a base level monitoring reading might be established, with no requirement for regular reading unless damage appears. At that point, the change in level should, by its detailed pattern, provide a powerful clue to its root cause.

Several specimens

The unpredictability of root growth means that, when several specimens surround a building, it is difficult to decide which of them are causing the damage. Tree root identification may condemn offending specimens (or at least confirm the species), but it cannot prove the innocence of the others. A level survey and level monitoring will usually indicate that one specimen is a more likely culprit than another. As already noted, after a repair-only remedy, the building can be monitored, enabling vegetation control to be reviewed. However, in most circumstances the underpinning option would have to assume that every specimen is doing (or will do) its worst.

Vegetation older than building

It is sometimes argued that the option of tree removal is precluded if the building is younger than the vegetation. This is based on the fear that recovery will continue above the horizontal axis (Figure 25.2, *page 258*) by an unknown amount, potentially causing worse damage than the original subsidence. However, that would be likely only in the following limited circumstances: the tree has already been causing a moisture deficit *below foundation level at the time the building was constructed.* Figure 25.5 (*overleaf*) illustrates the progress of subsidence when this condition does and does not apply. Building X, although younger than the tree, starts life beyond its influence; but building Y, also younger than the tree, is built, without precautions, within its influence. In time, both buildings suffer subsidence as the tree extends its root system, but only building Y goes on to suffer heave.

It would be useful to predict whether heave will follow tree removal in the circumstances depicted in Figure 25.5. This can be done, with reference yet again to the generic curve. Figure 25.6 (*page 265*) illustrates the case of X and Y on the familiar movement versus time graph. A building with no moisture deficit beneath it at the time of construction (X) is just another subsidence case. A building with a moisture deficit beneath it at the time of construction (Y) will eventually heave. That much is obvious. When it comes to prediction, the point is that a period of stability, in case X, is likely to precede onset of movement and damage. In case Y, movement and damage are likely to be apparent soon after construction. It follows that if a building has remained stable for some years in the presence of an older tree, the sudden onset of damage is likely to herald first-ever seasonal (stage B) subsidence, and the safest remedy would be tree removal. That would be the type-X case. (In actual cases, weather conditions during the stable years, and other evidence including distortion surveys, especially on the ground floor, can also be used as indicators.) For a type-Y case, we have the difficult task of estimating heave (Chapter 2). Nevertheless, doing nothing (in other words, neither controlling the vegetation nor underpinning the building) is not usually a realistic alternative. Mere pruning, as already discussed, will not be a permanent cure.

Stage removal

It is sometimes recommended that if a tree causing subsidence is removed, it should be done in stages – say, two rounds of reduction at yearly intervals

before the felling. The reasoning is probably that a slower recovery will
be gentler on the building. In practice, the effect of stage removal on rate
of recovery is usually overshadowed by other factors, such as soil permeability, weather conditions and, of course, how the tree responds to
the initial attacks on it. There is not much to lose by instant felling, and
something to be gained – an earlier end to recovery.

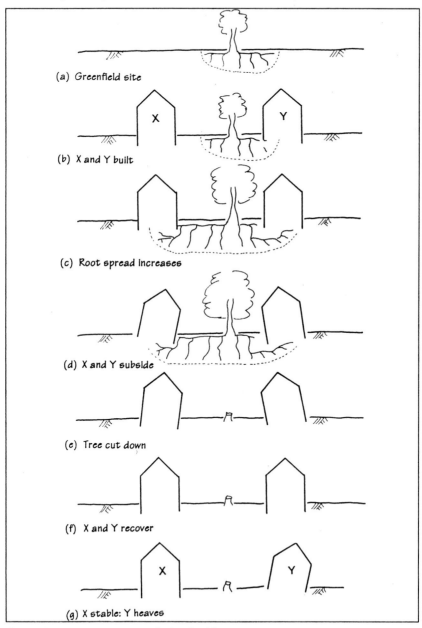

(a) Greenfield site

(b) X and Y built

(c) Root spread increases

(d) X and Y subside

(e) Tree cut down

(f) X and Y recover

(g) X stable: Y heaves

Figure 25.5 Vegetation predating building.

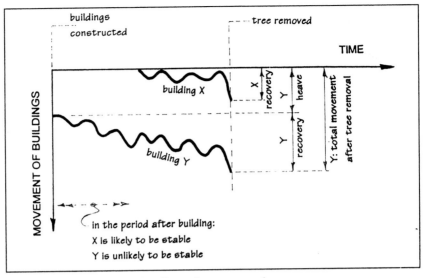

Figure 25.6 Predicting the effect of tree removal.

Destabilizing slopes

When clay loses water to vegetation, its cohesion is increased. Vegetation, therefore, improves the margin of safety of clay slopes (Chapter 3). Conversely, vegetation removal increases water content, reduces cohesion, and thereby reduces the margin of safety. Occasionally, the result can be instability (Figure 25.7, *overleaf*). The risk is greater with cuttings than with natural slopes, since the latter are more likely to have achieved stability without the help of vegetation. Inserting a barrier is not necessarily a safe alternative, because this can destabilize a slope by reducing the area of soil it needs to rely upon to resist disturbance (see B in Figure 3.12, *page 39*, and the accompanying text). A geotechnical engineer should be consulted where there is a potential for destabilizing a slope. Underpinning is sometimes the only safe option.

Heave

From the soil's point of view, heave is the same process as recovery; but heave is the preferred expression for movement fuelled by a moisture deficit present at the time of construction. The case of building Y (Figures 25.5 and 25.6) is covered by this definition. More common is the case where tree removal was carried out shortly before construction, and foundations were designed in ignorance of the hazard this introduced. In all

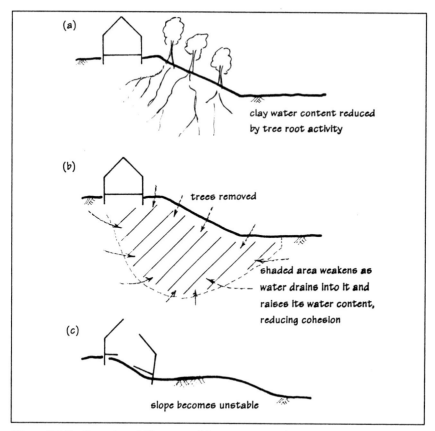

clay water content reduced
by tree root activity

trees removed

shaded area weakens as
water drains into it and
raises its water content,
reducing cohesion

slope becomes unstable

Figure 25.7   Slope instability caused by tree removal.

cases, the preferred remedial action is to let the heave work its way out, except if any of the following circumstances apply:

- Other causes also operate and demand additional attention.
- The building has become dangerous, or is likely to be endangered by the expected movement.
- The estimated permanent distortions will be unacceptable.
- The estimated time for heave to wane is unacceptably long.
- The cost of repeated repairs during a vigorous or prolonged heave is likely to be unacceptably high.

The decision is based, therefore, on an estimate of the severity of heave. This estimate is difficult to make; it is perilous to underestimate, because severe heave can be devastating, especially if it operates laterally as well as vertically. While acknowledging always the element of unpredictability, it is possible to judge likely severity from early progress. For example, a house that had developed 200mm of differential movement across the width of its lounge, within fifteen months of construction, was giving a clear signal that its heave was severe and action was imperative. Where there is doubt, level monitoring will measure progress. This should be disciplined, with frequent readings, urgent interpretation of results and two-way communication with the owner (who may be the first to perceive a worsening of the damage). Zero movement, even over a period of several months, should not be taken as proof that the heave has been spent, although it would in most circumstances suggest that subsequent movement should be less than severe, and underpinning would be unlikely to be necessary.

If underpinning is the chosen remedy, few circumstances will permit less than a full scheme, in which the entire ground floor and all foundations are provided with fresh support before being isolated from future vertical or lateral ground movements.

Subsidence

Erosion by water: granular soil and made ground

The options (Figure 25.8) are simpler than they are for clay subsidence. There is no recovery. The soil consolidates in the same way as made ground. In fact, reference to Table 9.1 (*page 81*) may help to predict residual subsidence after removal of cause. Some soil movement is inevitable. Eventual stability is also inevitable, but how long it takes will vary from case to case. The normal option is repair-only, which is preferred unless any of the following circumstances apply:

- Other causes also operate and demand additional attention.
- The building has become dangerous and needs urgent rescue.
- The estimated further consolidation would cause unacceptable damage.
- Removal of cause is not possible.

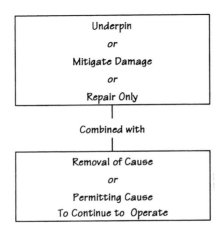

Figure 25.8
Granular or made ground subsidence: options

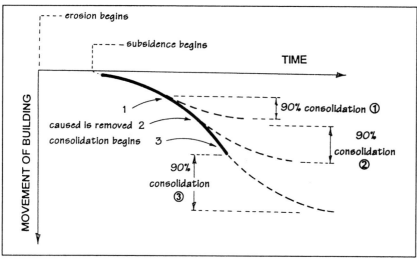

Figure 25.9 Removal of cause: timing – coarse soil.

If the option chosen includes removal of cause, the earlier this is done the better. Figure 25.9 makes the obvious point that delay increases the consolidation that will follow as well as the time taken by consolidation; and, of course, it considerably increases total movement.

Consolidation following removal of cause can be monitored, to confirm prediction or to assist in deciding when to carry out repairs. Results of the monitoring, as implied by Figure 25.9, should reflect a decreasing rate of subsidence until the threshold is reached where no damage occurs. Minor stop–go movement would, as always, superimpose itself on the exemplary curve, so monitoring should be continued for a few months beyond apparent stability.

If underpinning is the preferred option, it should be taken to a depth below unstable soil. On the few occasions when the cause will continue to operate, underpinning should also be taken deeper than any likely future instability. Where the cause is leakage from drains or mains, the area of loosened soil is often quite small in plan. Mitigation (Figure 25.10) can be an attractive option in those circumstances. (See also Chapter 17.)

Inundation: made ground

This is often referred to as 'collapse compression'. As explained in Chapter 9, it usually affects poorly compacted or exceptionally dry fill when water percolates downwards into it or ground water rises through it. There is no limit to how long made ground can remain stable but vulnerable before subsiding under inundation. Although it is a less common cause of subsidence than erosion, it is more difficult to remedy, for three reasons:

- The amount of subsidence is difficult to predict.
- It may not be possible to remove the cause (rising ground water, for example, is not easy to stem).
- Even if the cause can be removed, the subsidence does not necessarily stop soon afterwards.

The 'wait and repair' option is much less palatable than usual in these circumstances, especially as damage may reach an intolerable level during the inevitable delay, while geotechnical investigation and testing is carried out and interpreted. It would often be advisable to start preparing an underpinning design, as soon as the cause is understood in outline, so that this option can be applied at the earliest opportunity if it is found to be suitable. Underpinning would need to be protected from further collapse compression of the made ground. If the made ground is subsiding rapidly, difficult to penetrate by underpinning piles, or very thick, demolition may prove to be the only practicable option.

Landslip

Chapter 3 discusses in outline the causes of landslip (Figure 3.11, *page 38*). It is a specialist subject, and further technical discussion would be outside the scope of this book. It is worth emphasizing again that early diagnosis of landslip is even more important than it is for other insured perils, because mass movement is not easily resisted by foundations. Neither can underpinning usually resist it. In most cases of landslip, the

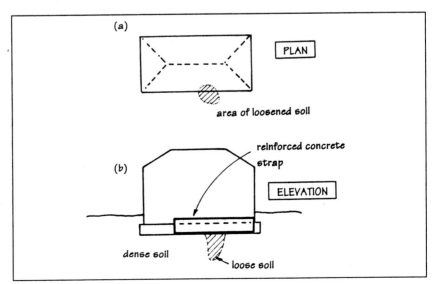

Figure 25.10   Bridging across loosened soil.

options narrow down to different means of stabilizing the slope before doing anything to the building. Placing a gabion or similar massive wall (Figure 3.12D, *page 39*) on firm foundations is probably the most common remedy. Sometimes this is not practicable and sheet piles (Figure 3.12E) must be used.

Management of progressive damage

The most useful skills are good timing and communication. Both cost and inconvenience are very much reduced by:

- early diagnosis
- early removal of cause
- rapid production of any additional technical information
- appropriate timing of repairs.

Early diagnosis

Subsidence and landslip are likely to escalate and cause increasing damage until action is taken to arrest or bypass the cause. Obviously, no action can be taken until at least a preliminary diagnosis has been made. There are few occasions when diagnosis cannot be made quickly, provided funds are available for sufficient investigation. For most cases, a visual inspection (by the diagnostician, not a reporter; see Chapter 7), level survey and trial hole are sufficient. More intractable problems require more sophisticated information, but there need not be any significant delay in commissioning and obtaining it. Monitoring introduces open-ended delay, and should not be used as a tool for diagnosis, except as a very last resort.

In the case of severe heave, underpinning is one of the remedial options, and late diagnosis would allow damage to accrue that underpinning could and should have avoided.

Early removal of cause

Except in the case of heave, the majority of insurance claims are solved by removing the cause. The case for early removal is the same as for early diagnosis. Even if it is decided later that removing the cause was not enough, some benefit will have been gained by slowing down the progress of movement and damage. So, even when there is doubt about the final remedy, it is usually worth removing the cause of damage as soon as it has been identified – provided, of course, that it is both practicable and legal to do so.

Additional technical information

In some cases, a third party has to agree to removing the cause of damage, and may need to be convinced of the need to do so before giving assent. That may, in turn, call for a higher level of proof than the repairer would need simply to specify repairs, a circumstance that is fertile ground for disputes.

Trees are the centre of many such disputes. Not infrequently, a prolonged impasse develops between the insured's advisors, who advocate tree removal on the strength of a straightforward diagnosis, and the tree's owner as well as the local planning authority, if there is a tree preservation order (TPO). (In urban areas, the local planning authority may be both the owner and preserver.) TPOs prevent felling or any form of reduction without local authority permission, which is not granted carelessly. While the impasse prevails, damage escalates and the alternatives for repair can dwindle to a single expensive option.

Whenever possible, potential disputes should be anticipated at the outset. Resources should be made available as soon as possible to enable investigators to obtain causation evidence of a quality that would be difficult to refute. Clearly, proof cannot be guaranteed in advance. It may turn out that the additional tasks do not provide sufficiently powerful indicators to convince those who need convincing. It is better to arrive at that point sooner rather than later.

The additional tasks required to support a case for tree removal normally include positive root identification, which may require energetic searching. Even dense root growth can elude two or three 100mm diameter boreholes. Wide trial holes are more likely to find the evidence, if it is there. Monitoring should start as soon as possible, and readings should be frequent enough to register all significant movement. If monitoring misses any peaks or troughs, it will understate the case. Distortion surveys and soil testing should be comprehensive enough to make it difficult to advocate more than one interpretation.

If the issue of liability does not affect the removal of cause, gathering additional information need not delay the progress of the claim.

On the rare occasions when monitoring is needed for diagnosis, it should be initiated at the earliest opportunity. More often, monitoring is used to confirm that removing the cause was the right decision. It is also used for timing repairs. Action cannot be taken until enough results have been obtained so, again, early installation is advisable. The first two or three readings are often the most informative. In the case of heave, early readings are crucial if monitoring is the means of deciding whether the case is severe enough to warrant underpinning, and it would be difficult to make a plausible excuse for delay.

Timing of repairs

It is safe to carry out repairs when the rate of recovery or heave falls below the threshold of damage (Chapter 9 and Figure 9.5, *page 85*). This moment is unpredictable and unrecognizable, except after a long enough period of hindsight. Only litigators can afford plenty of hindsight. Owners are usually anxious to enjoy full use of the building before stability has been proved beyond doubt. So a compromise must be sought. It may, for example, be agreed to start repairs *when the risk of cosmetic damage arising from future movement has become low and the risk of serious damage has become very low*. The possibility of a further, less expensive, round of repairs then has to be accepted as the price of uncertainty. (These comments apply to recovery or residual heave on clay, and to residual subsidence on granular soil or made ground.)

The timing of repairs is also a consideration after underpinning. The same principles apply. This time, the movement to be avoided or alleviated is settlement of the building on its new foundations. It is unusual for such settlement to be either great or prolonged. More often than not, the underpinning has by its nature a wider margin of safety than is typical of new foundations. Even if repairs take place before the consolidation phase is complete (Figure 9.1, *page 80*), the repaired building has to cope with millimetres rather than centimetres of movement. In the case of piles that rely mainly on shaft friction (not end bearing) for their working capacity, it is often difficult to measure any settlement at all after completion of the underpinning contract. The benefits of completing remedial work without a pause for 'bedding down' (during which the building may remain unused and probably damp) outweigh the risk of minor damage, except for those few cases where the soil supporting the underpinning is less than dense or stiff. Jacking (Chapter 17) may reduce the waiting time in such cases.

Communication

All the parties and advisors need to understand:

- the cause of damage
- the options for remedial work
- what is going on at the moment.

During monitoring, there are periods when very little is going on. Owners in particular welcome regular contact during these periods and assurance that nothing is being forgotten.

Remedial works contract

The contract for the remedial works is a simple one between the owner and the contractor, managed by an agent who acts as administrator and authorizer of funds. Sometimes there are two agents: a building professional responsible for day-to-day administration and a loss adjuster responsible for the release of funds. One of the dangers of this arrangement is the ordering, by the professional, of variations later perceived by the loss adjuster as outside the terms of the claim, so that the owner is faced with an unexpected bill.

Non-progressive insured perils

Damage caused by wind, fire, explosion or impact can be regarded as a single incidents (Chapter 16), and repair is usually simple. Part Three contains typical repair details. Floods do not usually cause structural damage unless:

- the soil is vulnerable to inundation (Chapter 9; Figures 9.3 and 9.4, *pages 82 and 84*)
- the structural material is vulnerable to inundation (Chapter 15)
- impact damage has been caused by the force of water or the debris it carried.

It may be necessary, after flooding, to monitor the building for rot (Chapter 15) and shrinkage (Chapter 12) during the drying-out period.

Ombudsman

The domestic (household) insurance policy is a contract between owner and insurer. In the event of a dispute, the courts are willing to make a judgement based on contract law, but it is much better to refer to the ombudsman first. The ombudsman procedure is inquisitorial rather than adversarial, allowing more imaginative solutions to be achieved than are possible through litigation. There is also a much better chance of reaching an amicable outcome. Sometimes incomplete information is part of the reason for the dispute. The ombudsman has the power to commission an additional investigation, if that is likely to be useful. This offers the prospect of an appropriate solution, in preference to a stark choice, based on limited information, between two intransigent viewpoints.

Chapter Twenty Six

Law

This chapter lists a few aspects of law that have a daily impact on structural repair. Some construction law firms produce newsletters and seminars, which keep their clients up to date with important changes.

Tort

Any firm or person can successfully sue a building professional or firm, if it can be proved that:

- The professional owed them a duty of care.
- The professional was in breach of that duty.
- The firm or person suffered damage as a result of that breach.
- The damage was discovered within a set period of time following the breach, and the writ was also served in time. (The periods are defined in the Limitation Act 1980.)

Building professionals are liable in tort for negligence in their investigation, design, supervision, management – or indeed in any advice they give. Consultants may feel pleased (or may have felt pleased at the time) with the way they were able to adjust their design to unforeseeable conditions. Their client, who had to pay for the adjustments, may question whether the conditions were really unforeseeable or merely unforeseen. A second opinion may be eager to suggest the latter. The consultants' disquiet begins when they review decisions they made with foresight in the harsher glare of hindsight, and they find that what seemed pragmatic and reasonable at the time could be reinterpreted by the ungenerous as arbitrary and careless. The courts are advised by experts in the subjects under scrutiny, but it is not difficult to find experts who will disagree on what a reasonable professional should have done at the time. Their divergence of opinion keeps disputes alive. Table 26.1 lists some of the more common technical errors that lead to litigation.

Table 26.1 Tort: examples

Technical mistakes
- Working outside one's area of competence.
- Being unaware of new knowledge.
- Neglecting advice in codes and approved guidance, such as the principles of prevention and protection (Appendix B).
- Serious errors in structural analysis (less common than other matters).

Organisational mistakes
- Failure to absorb changing technical or legal requirements into standing advice.
- Poor training and briefing of key personnel.
- Appointing the wrong people to important roles.

Contract

Whether or not a contract is formal, or even recorded in writing, its essential ingredients are:

- an agreement between the parties
- a consideration
- the intention of the parties to be bound by their agreement.

When repairers enter into a contract with their client, the agreement is normally for them to prepare a design (and probably carry out site inspections). The consideration is normally their fee. Serious misunderstanding is possible unless both are carefully defined. Table 26.2 lists some of the more common breaches of contract committed by building professionals.

Table 26.2 Breaches of contract: examples

- Failure to define the limits of agreement.
  (e.g. A full service expected but design only provided.)
- Omitting part of clients' instructions.
  (Which may include important non technical elements.)
- Presenting incomplete work.
  (e.g. Drawings not detailed enough for construction; reports limited in extent.)
- Inadequate inspection of contractors' work.
  (Failure to visit at certain stages; inadequate understanding of quality).
- Inadequacy of sub-consultants' work or information.

Repairers may also act as agent to the contract between the client and the contractors carrying out the repair. The daily administration of routine building contracts can breed a false sense of ability, and repairers can be caught out when disputes arise. Two of the most common mistakes are incorrectly valuing work and failing to condemn work of poor quality. The latter could provoke a claim in both tort and contract.

Health and safety law

All workers, from the tea boy to the contractor's managing director, have a duty to look after their own health and safety and to safeguard others who may be affected by their actions. The level of responsibility varies from person to person, according to their knowledge and their role in the design and management of the work. Repairers are on duty, as far as health and safety are concerned, throughout the entire project – investigation, design and inspection of work. They remain liable afterwards, for any breach that injures people who use the building, or clean or maintain it.

An accident may appear to be the result of a moment's carelessness, but any enquiry or prosecution will want to trace back to events or failures that led up to the moment, or perhaps shaped it: company policy, site rules, health and safety plan, inspection of work and design. It can often be demonstrated (Chapter 6) that shortcomings in the design were contributory factors to the accident. In that case, designers may be prosecuted under the Health and Safety at Work etc. Act 1974 and the Construction (Design and Management) Regulations 1994.

Repairers should have a general knowledge of the main health and safety acts and regulations (Table 6.2, *page 57*) and a good understanding of those that impinge on their everyday work.

Planning

Repairers should also be aware of the law as it applies to historic buildings. It is best to assume that no historic building, and no building within a conservation area, may be touched without permission. Even taking up floorboards to inspect a beam can prove to be an offence unless authorized by the local planning authority (Chapter 5). PPG 15 and PPG 16 (see *Further reading*) provide general guidance. On specific matters, advice is usually readily available from the local planning authority or English Heritage.

The Party Wall etc. Act 1996

Repairs to a party wall or any other shared structure, and any excavations carried out near to an adjoining building, come under the Party Wall etc. Act, whose intentions are to deter thoughtless work on shared structural elements, thereby reducing unnecessary damage and blight. The Act has the distinction of being only fifteen pages long and fairly easy for the non-expert to understand.

Owners proposing to do work on any building element covered by the Act (the Act contains clear definitions) must give notice to adjoining owners. The work may not go ahead until there is a positive agreement, and not just silent acquiescence. Failing agreement, a binding 'award' has to be made by a party wall surveyor (or surveyors).

It is also necessary to serve notice on adjoining owners of a proposed excavation, if it falls within either the 'three metre rule' or the 'six metre rule'. This requirement is illustrated on Figure 26.1. It covers not just

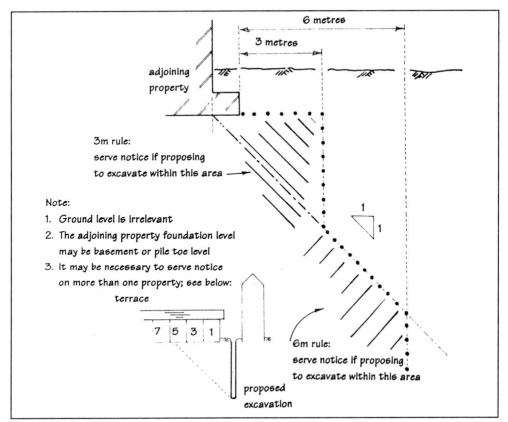

Figure 26.1   Party walls: three metre and six metre rules.

underpinning and other below-ground work, but drainage trenches and even simple trial holes.

As well as providing a procedure (and timetable) for serving notices and managing any disputes, the Act confers rights on those proposing to do work and on adjoining owners who may be affected by it. For example, building owners or their agents are allowed to enter land for the purpose of carrying out work permitted by the Act.

Adjudication

Until the Housing Grants, Construction and Regeneration Act 1996 came into force, any valid and legal contract was self-contained and immune to outside interference. Now, any party to a construction contract has the right to refer a dispute arising under the contract for adjudication under an agreed procedure. There is a timetable for this procedure. The dispute must arise 'under the contract' and cannot include allegations of negligence. Most construction contracts are covered by the Act, including those between clients and their professional advisors, with the exception that house owners are not bound by it. The Construction (Design and Management) Regulations 1994 (CDM) are also not binding on house owners.

One of the main purposes of the Housing Grants, Construction and Regeneration Act 1996 is to reduce payment disputes.

Insurance

Building professionals should be covered by professional indemnity insurance, which is a contract based on utmost good faith. This means, for example, that the insured is obliged to provide all relevant information to the insurers and must comply with all the insurers' requirements regarding the management of any claims that arise. Early notification by insured to insurers is usually considered essential. Most insurers take on the management of claims against their insured.

Building insurance policies are also contracts. Their interpretation by claims officers is flavoured by commercial practice, and departures from strict legal interpretation of the policy usually favour the insured. As noted in Chapter 25, disputes can be referred to the ombudsman.

Alternative dispute resolution

The insurance ombudsman service is an example of alternative dispute resolution (ADR), whose general purpose is to settle disputes more quickly and with less cost than litigation. ADR also has the advantage that the parties retain control over the process, instead of handing it over to the court. There is obviously the corresponding disadvantage that one of the parties may decide to abandon the process before a binding result is agreed.

There are more types of dispute resolution than there is space here to list and define. Table 26.3 includes the more common types. Roughly speaking, the higher up the table, the more expensive the method, and the less control the parties have once the process has been launched. Arbitration is a legal process run by an appointed arbitrator, and it cannot be abandoned except by a court order or, of course, the agreement of the parties to settle before the award is made. At the other extreme, mediation is usually run as a form of shuttle diplomacy; the mediator acts as a catalyst to negotiations, without expressing personal opinion. This method has to be abandoned unless mutual agreement is reached and then, preferably, put into writing and signed. ADR can be a two-stage process. Two-stage resolutions start at the lower end of the table; if not successful at that level, they pass on to another method, so that early failure does not necessarily prompt litigation. For example, if mediation is the first stage,

Table 26.3   Alternative dispute resolution: examples

Litigation	
Arbitration	
Expert determination	*An expert appointed by the parties makes a binding decision.*
Adjudication	*Determination by an adjudicator whose decision may or may not be binding according to the terms of the agreement.*
Mini-trial	*A conference where evidence is presented in short time and appraised by a panel of disputants, or a neutral advisor.*
Ombudsman service (*domestic insurance*)	*The Ombudsman conducts an inquisitorial procedure, and gives an opinion not binding on insured but binding on insurers.*
Conciliation	*A structured negotiation with the conciliator expressing an opinion or giving a non binding decision.*
Mediation	*Negotiation assisted by a neutral person.*
Direct negotiation	*Followed by an agreement to be bound permanently, or temporarily – say until the end of the contract.*

but no mutual agreement is reached, the mediator may then evolve into a conciliator, if the parties wish.

Managing liability

Liability can be limited by identifying the risks, managing them, evaluating one's success in doing so, updating policy and making sure that staff are educated and informed.

Managers have a survival instinct that warns them of severe risks; but it is difficult to identify all the risks and rank them correctly, without instigating a formal appraisal based on:

- a review of typical work instructions
- analysis of serious or recurring problems
- an appreciation of construction law.

For example, the appraisal may identify the firm's core 'bread and butter' work, and subdivide it into a number of tasks, so that a large percentage of technical work falls within a list of defined activities. Some of them can be highlighted for their above-average technical difficulties, some for their tendency to breed delays, their vulnerability to upsets caused by outside firms, and so on. These are the risks most likely to benefit from management. Policies may then be evolved for controlling them. Outside the core activities, unusual instructions are prone to errors through inexperience, and someone should be on hand to anticipate such errors. They may be awkward to manage by formal procedure, but they can be tagged as perhaps needing more than average supervision by a senior person.

All building professionals and firms have unique strengths and weaknesses, and a unique pattern of work. Their risk management should reflect this. Most schemes would nevertheless share many of the following features:

- Risk management should not be an isolated discipline. It works better if incorporated into general management, including quality assurance and health and safety management.
- The start of a job is the most fruitful time to reduce risks, not only by identifying where they are but, for example, deciding who would be best at managing them.
- Recognized (checkable) methods should be used to deal with uncertainty. A second opinion may be invited on crucial events.
- Technical sensitivities should be recognized. Unforeseen conditions are the sensitivities that haunt most consultants, but others are technical and organizational. (Examples are: changes to a building causing instability;

changes in a project team undermining confidence or expertise; critical events whose delay is always difficult to recover; and open-ended delays caused by other organizations, including those with a duty to determine applications for approval.)

- Each job should be reviewed at intervals so that action is taken to hold down the expected risks and identify new ones.

Since risk cannot be eliminated, a few mishaps will survive the greatest of schemes. It is much easier to defend one's actions against unjust criticism if a record has been made of important decisions and the principles that were applied to them. Standard forms for giving and receiving instructions, serving notices, minuting meetings and phone calls, and so on, are helpful prompts in themselves. However, the people using them must be aware of when it is necessary to depart from the standard, and when it is prudent to be suspicious of others urging them to depart from the standard.

Quality schemes normally record complaints. The analysis of complaints helps to detect risk management failures, thereby contributing to an evaluation of the scheme, and a discussion of possible improvements. This is a policy that works best in an atmosphere of trust rather than suspicion or anxiety.

Complaints are not the only spur to evaluating and improving the scheme. Changing demands and conditions must be anticipated. Regular clients may suddenly wish to impose new conditions, for better or (more often) worse, in the name of efficiency. The national economy has a medium-term influence on the market and on commercial policy. It is well known that legal disputes increase during a recession, and diminish during times of prosperity, when there are other opportunities for income.

Every significant change to the risk management scheme provides an opportunity to inform and educate the people who work it. Education provides the opportunity to remind everyone that risk management is not restricted to following and recording set procedures. Personal skills are as important. When things are not going well, the good manager is the first to detect the client's (or other team members') growing irritation with small problems, and the drift towards increasingly formal relations. There is a time when the causes of these symptoms can still be addressed, before they ripen into conflict.

If, unfortunately, the differences seem to be lumbering beyond tactful intervention and towards a formal dispute, the professional indemnity insurers should be notified immediately. It may not be too late to resolve difficulties, if the insurers agree, by negotiation or ADR. It is always essential to seek early legal advice, which may well be provided by, or arranged by, the insurers.

Chapter Twenty Seven

Surprises

Hidden details

In any building, only part of the structure is on display. Routine appraisals have to made without knowing what all the materials are. The infill of a medieval timber-frame building may consist of wattle and daub in some places, brickwork in others, earth in others, all covered by render and plaster. Any contribution the infill is making to the building's robustness, which may be more than useful if the frame has been altered, must be judged without knowing what it is made of. The older the building, the more difficult are these problems of identification, although there are modern examples. For example, a rustic bungalow in Suffolk, with well-maintained facings and decorations, yielded few clues at ground level to its structural origin. Fortunately, a brief look into the roof space revealed, below the low pitched rafters, the tops of two twentieth century railway carriages side by side.

Several references have been made in previous chapters to hidden details (for example, bonding timber in masonry, Figures 15.7 and 19.14, *pages 143 and 196*). If hidden details are impairing structural performance, they will provide clues to their existence, although not always to their where-abouts. Figure 27.1 (*overleaf*) shows an example of damage that would be very difficult to trace to its origin, especially at its first appearance. No one can diagnose nascent damage. The cost of investigating the causes of trivial symptoms can rarely be justified. If they are not causing pro-gressive damage, hidden imperfections remain hidden unless accidentally discovered when coverings are stripped. All too often this happens when scheduled repairs are being carried out; the revelations then become the subject of variations, an irritant to many well-planned contracts.

The most difficult and expensive problems often lie below ground level. This applies to hidden imperfections as well as to the familiar causes of damage. Among the less welcome surprises are old wells and pockets of domestic or commercial waste (sometimes excavated for this purpose, sometimes providing a new use for old ditches). Examples of unwelcome materials – unusual but not unique – have included:

- unrecorded burial pits for farm animals
- dumps of banned products such as toxic chemicals and asbestos
- abandoned mine shafts whose 'filling' proves to be impermanent
- stolen property.

Sometimes the material is unexpectedly uncovered by preparations for repairs; only then is it found to have been contributing to the movement and damage that prompted the repairs in progress. The structural design has to be amended urgently and, if carcasses or chemicals are discovered, health and safety precautions must be reconsidered before any further work is done. It is then that even the most experienced investigator rue-fully accepts that nobody's knowledge of building defects is complete.

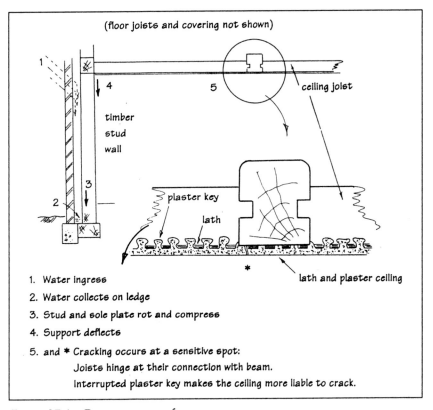

1. Water ingress
2. Water collects on ledge
3. Stud and sole plate rot and compress
4. Support deflects
5. and * Cracking occurs at a sensitive spot:
   Joists hinge at their connection with beam.
   Interrupted plaster key makes the ceiling more liable to crack.

Figure 27.1   Damage remote from cause.

Concealed defects

Occasionally, defects within the structure are cunningly disguised by owners without the funds, or the will, to deal with them when they become manifest. The offenders hope to be elsewhere before the defects reappear. Surveyors have no such choice. With or without the aid of symptoms, they must pass judgement on the condition of the building. With luck, their suspicion might be raised by unusual decorations: fussy beading around openings (disguising their mis-shape); oversize skirtings, new ceilings or dry lining in just one or two rooms (disguising gross distortion of walls or floors); or any *ad hoc* bracing, strutting or tying that might represent a token response to local decline. Even so, without symptoms on the surface, it is not usually possible to justify destructive investigation. Unfortunately, by the time symptoms penetrate the camouflage, the building may be in poor shape.

Geometry

The older the building, the more time there has been to carry out strange alterations. Sometimes, the only way to establish accurately superimposed floor layouts is to hang three plumb bobs from roof to basement, surveying each floor relative to the triangle they form. On one occasion, this showed that the party wall and a large chimney were seriously offset at successive levels, and stability was relying precariously on a very few unassuming floor joists. The proposed modest alteration to the ground floor restaurant might have been the least, but possibly the last, of many alterations.

Geography and history

Except for the most prestigious buildings, construction before the twentieth century was dominated by the cost of transport. Local materials were preferred (timber in East Anglia; limestone in West Yorkshire) except where a well-established trade route was open between supply and demand. Building techniques changed over the years, as did the layouts of buildings and the fashion in decorative styles. Thus, materials and methods are characteristic of their time and place. Dating buildings is a useful clue to their structure (Table 27.1), although the non-expert can be deceived by later alteration, which may present more vivid evidence than the original parts. Even a date stone sometimes merely dates the stone. It may only be part of a local repair. Documentary evidence

Table 27.1   Construction dates

1500	1600	1700	1800	1900	2000

EARTH

TIMBER

Use of jetties

Oak used in main frame

Elm

Imported softwood

Size and quality of structural members reduced; infill contributes to structural performance

Modern timber frame

External render

MASONRY   Brick tax introduced   Brick tax removed

Solid walling

Cavity walling

Lime mortar

Cement mortar

Bonding timber

Hoop irons   Expanded metal

Bed course metal reinforcement

Predominant type of wall tie

Vitrified Clay   Cast Iron   Steel   Stainless Steel

Increasing use of non traditional floors

Increasing use of modern rigid structural frames

1500	1600	1700	1800	1900	2000

Peak usage is shown by solid line. There are regional variations in material availability and construction history. Many buildings contain reused material. Alternatives are not mutually exclusive e.g. the same building may contain lime mortar in the internal (or lowly stressed) masonry but cement mortar in the external (or highly stressed) masonry.

(maps, directories, census) is sometimes available from county records. Sampling (if permitted) and testing may produce vital information. Dendrochronology, for example, can usually date the felling of native timber to the exact year through the study of their growth rings.

Cast iron appeared at the end of the eighteenth century. From then on, many new materials were introduced (Table 27.2) into otherwise traditional buildings. During the twentieth century, increasing use of rigid frames and off-site production has reduced the proportion of building that can be described as traditional.

Table 27.2   Approximate date of introduction of materials

1790	Cast iron
1840	Wrought iron
1860	Filler joist floors (iron, later steel, joists in concrete)
1870	Machine moulding of bricks First appearance of flettons (commons) and perforated units
1875	Dense concrete blocks
1885	Steel
1900	Precast concrete floors
1905	Calcium silicate brickwork
1918	High alumina cement (until late 1970s)
1920	Heavy-duty timber connectors
1920	Concrete panels and blocks as facing masonry
1923	Fletton bricks as facings
1930	Hollow clay tiles as backing masonry
1930	Resin glue
1936	Lightweight concrete (made with lightweight aggregate)
1950	Prestressed concrete floors
1957	Lightweight concrete (air entrained)
1965	Trussed rafters
1970	Prestressed brickwork
1981	Stainless steel wall ties

Note: Occasional examples (imports, early prototypes etc.) appeared in the UK before the dates given.

Government policy

Governments have occasionally influenced construction by taxing materials and features. For example, the tax on bricks encouraged the use of weak material in East Anglia and other parts of the country. The use of weak materials declined when the tax was removed in 1850, but it continued into the early twentieth century and, on a very small scale, even today. There are few abrupt changes in material usage. One of the few came immediately after World War II, when the government strongly encouraged the use of factory-made panels and frames, and precast buildings enjoyed a brief popularity.

Gradual and sudden failure

Failure can, very occasionally, be sudden (Figure 20.1, *page 198*), even when defects have not been concealed. It is impossible to provide a list of circumstances favourable to sudden failure; hence it is impossible to guarantee that sudden failure will be anticipated by inspection.

As a very rough rule, failures caused by bending are gradual. Examples are overloaded floors and in-plane movement of masonry walls, such as sagging and hogging (Figure 20.2a–d, *page 201*). Sudden failure can be caused by loss of equilibrium (Figure 11.1, *page 108*), out of plane bending of masonry walls and high shear stresses. Figure 27.2 shows two examples of timber weakened by rot. In case (a), the walls of a sports hall consisted of studwork enclosed by weatherboarding and plasterboard. Rot developed in the studwork at a level where damp had entered through a poor detail, but there was no evidence of deterioration on either face of the wall. The owners of the building arranged for an inspection when they noticed that the walls had begun to shiver in the wind. Had the rot persisted, the studwork would have failed in bending but, because bending failure was as usual preceded by serious distortion, it gave notice of itself.

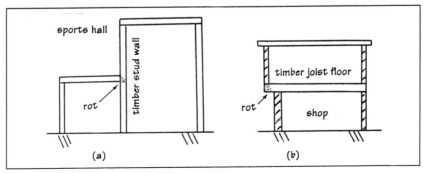

Figure 27.2   Gradual and sudden failure.

The owners of the sports hall were creditably alert to symptoms that developed by small increments. The tenants of the shop (b) had no chance to anticipate the sudden failure of its stub cantilever joists, which brought down part of the storey above. Again the underlying cause was rot, but this time the failure was in shear, whose prior distortion would have been too small for even an expert to detect. The only warning was a flurry of plaster flakes a few seconds before the collapse, which at least gave shop window browsers enough time to run away and escape injury.

Success and failure

Design standards and codes are based on past failures. Failure prompts enquiry, whose purpose is to establish cause. When this proves to be unusual or not well understood by the industry, it promotes research, which eventually feeds back into standards and codes so that future design is better informed.

The advice given in British Standards and European Codes is state of the art – representing current theory and practice in achieving acceptable performance in new buildings – but they only face one way. It is not their function to explain or improve unacceptable performance in existing buildings. No equivalent codes are available for this; hence the need for investigators and repairers to have a sound knowledge of defects and material behaviour, and an appreciation that, with existing buildings, knowledge is always incomplete. Surprise has to be expected.

APPENDICES

Appendix A

Risk assessment: health and safety

Repairers with limited site experience would find the Construction Industry Training Board's *Construction Site Safety – Safety Notes* (GE700) a comprehensive introduction to hazards, written in plain English. HSE and CIRIA also provide many useful publications.

Hazard is the potential to cause harm. Severity is the level of harm. Likelihood is the probability that the harm will occur.

Risk = severity × likelihood.

More than one method of risk assessment is in circulation. One of the simplest is a four-step procedure:

- Identify every hazard.
- Rate its severity (1, 2 or 3).
- Rate its likelihood (1, 2 or 3).
- Risk is then severity times likelihood, leading to a value between 1 (trivial) and 9 (extremely high).

Low levels of risk can be left to site management. High levels should be avoided or reduced by design. The cost of avoiding or reducing risk should be taken into account when making such judgements (Chapter 6). A moderate risk may have to be tolerated if the alternative were to add considerable cost and delay to the job. On the other hand, even low levels can be addressed if the cost of doing so is small.

A simple example of the use of risk assessment might be based on the underpinning of a fragile building with shallow footings. Continuous strip underpinning, by undermining the building throughout the contract, would attract a high severity and a high likelihood. Bored pile underpinning would be a lower risk and would therefore be preferable, unless it was significantly more difficult or expensive. If, for whatever reason, hand excavation remained the preferred option, after this initial consideration, risk assessment and control might be further informed by breaking down the work into several separate hazards, such as:

- variable (or unknown) footing profile
- condition of the building and its footing
- underground services or obstructions
- variable (or unknown) soil conditions
- variable groundwater level
- location of excavation (with respect to the site layout)
- the need for access around the site (especially vehicular)
- the need to store materials
- possible work congestion
- quality of shoring.

A what-if analysis can be applied to each: What if the footing is very shallow? What if it is defective? What if the wall it supports is slender/out of plumb? and so on. This builds up a risk profile.

In some cases, sensitivity may influence risk. In many soils, variable groundwater (from the list above) would introduce a need for standby pumps on site to control possible flooding, a nuisance but not a critical problem. On soils with a high proportion of fine sand or silt, a rise in groundwater could create sudden instability of the floor of the excavation. That would be less acceptable, justifying spending more on avoidance.

Finally, in some cases, the hazard may be a trigger that does not lead immediately to the unwanted target event. Likelihood, then, depends on whether enough unfavourable factors are in place between trigger and target. The sequence can be represented as a flow chart or event tree (Figure A.1). Event trees are useful tools for establishing likelihood and, by highlighting opportunities to intervene at critical points, they can also serve the principles of prevention and protection.

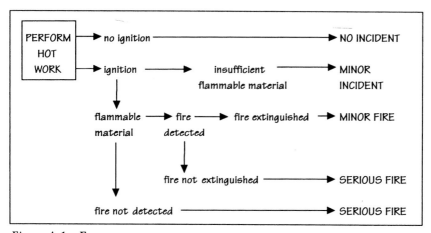

Figure A.1   Event tree.

Appendix B

The principles of
prevention and protection

The principles of prevention and protection are:

(a) *If possible, avoid the risk completely*, by using alternative methods or materials.

(b) *Combat risks at source*, rather than by measures which leave the risk in place but attempt to prevent contact with the risk.

(c) *Wherever possible, adapt work to the individual*, particularly in the choice of work equipment and methods of work. This will make work less monotonous and improve concentration, and reduce the temptation to improvise equipment and methods.

(d) *Take advantage of technological progress*, which often offers opportunities for safer and more efficient working methods.

(e) *Incorporate the prevention measures into a coherent plan* to reduce progressively those risks which cannot altogether be avoided and which takes into account working conditions, organizational factors, the working environment and social factors. On individual projects, the health and safety plan (Regulation 15 [of the CDM Regulations]) will act as the focus for bringing together and co-ordinating the individual policies of everyone involved. Where an employer is required under section 2(3) of the Health and Safety at Work etc. Act 1974 to have a health and safety policy, this should be prepared and applied by reference to these principles.

(f) *Give priority to those measures which protect the whole workforce or activity* and so yield the greatest benefit, i.e. give collective protective measures, such as suitable working platforms with edge protection, priority over individual measures, such as safety harnesses.

(g) *Employees and the self-employed need to understand what they need to do*, e.g. by training, instruction, and communication of plans and risk assessments.

(h) *The existence of an active safety culture affecting the organizations responsible for developing and executing the project needs to be assured.*

Extracted from 'Managing Construction for Health and Safety' published by HSE as the Approved Code of Practice for the CDM Regulations. (Crown copyright material is reproduced with the permission of the Controller of Her Majesty's Stationery Office.)

Glossary

Wherever possible, terms that are not self-evident have been explained within the text. The following have not. They do not have standard or approved definitions, but it is hoped that the following notes will be useful.

Soil volume reduction

Several terms are used to describe the process of soil volume reduction, and it is sometimes important to distinguish one from another.

Consolidation consists of one (or both) of the following events:

- expulsion of pore water pressure
- compression of the soil skeleton.

In fine soil, both events occur. In coarse soil, volume reduction arises from compression of the soil skeleton alone.
   Consolidation is caused by pressure, arising from any of the following:

- the weight of a new building or civil engineering structure
- soil or other material stored on the surface
- in the case of loosely tipped made ground, its own weight.

Collapse compression is consolidation caused by inundation of the soil. It affects mainly loose, dry made ground, not natural soil. The made ground usually consists of large particles, or aggregations of particles, and it compresses when:

- aggregations are softened (fine grained material)
- particles are weakened (coarse grained material)
- bond or friction between particles and aggregations of particles is weakened.

Compaction achieves volume reduction by mechanical means. It can be applied to made ground as it is tipped, or to made ground after it has been tipped, or to loose natural soil.

When applied to made ground as it is tipped, the purpose of compaction is to create strong soil that will not be prone to large subsequent consolidation. This is usually effected by placing the material in thin layers, each subjected temporarily to high pressure or vibration or both. A range of equipment is available. The choice depends mainly on type of material.

Made ground that has been tipped without compaction, and loose natural soil, may be compacted *in situ* by several means. These include:

- static pressure (temporarily placing additional soil on it)
- dynamic pressure (dropping a heavy weight onto its surface)
- displacement (driving piles into it)
- vibro methods (vibrating columns of the soil)
- various other means.

Each is suitable for a limited range of soil types and environmental conditions.

Settlement is caused by consolidation of the soil arising from an applied pressure *other than the soil's own weight*. In most cases, the pressure is applied by a new building; settlement starts as construction begins, and it finishes (in most cases) a relatively short time afterwards (Figure 9.1, *page 80*). Settlement may start again if the building is altered in a way that increases bearing pressure on the soil. It can also start again if a new building is constructed in the near vicinity (Figure 9.17, *page 102*), or if another load, such as storage or fill material, is applied nearby.

Subsidence is caused by consolidation of the soil under the influence of factors *other than externally applied pressure*. The most common factors are:

- shrinkage of clay caused by suction (Figure 3.5, *page 28*).
- consolidation of made ground under self-weight (Table 9.1, *page 81*).
- loosening of soil by erosion (Figure 9.2, *page 82*).
- collapse compression caused by inundation (Figures 9.3 and 9.4, *pages 82 and 84*).

Subsidence may also be caused by ground water lowering (Figure 9.9, *page 94*), solution (Figure 9.12, *page 96*) and mining (Figure 9.13, *page 98*).

Settlement and subsidence

Most cases of volume reduction fall within the definition of *either* settlement *or* subsidence, but occasionally the two may act at the same time.

For example, soil loosened by erosion and still subsiding as a result of that cause, may also settle if it is built upon, as shown by the dotted line in Figure 9.2 (*page 82*). Total volume reduction would be the addition of subsidence and settlement.

Differential movement

When downward movement is uniform, little or no damage is caused to buildings. It is *differential* settlement and *differential* subsidence that cause structural damage. The level of damage depends on how pressure and soil conditions vary and on how the structure copes with these variations (Figures 4.1, 4.2, and 4.3 , *pages 44, 45 and 46*).

Monitoring

Monitoring is the regular measurement of a building's movement or damage. The most frequently measured movement is vertical foundation movement, usually done by means of level readings (Figure G.1). A stable datum is set up, usually a rod anchored into the ground at a depth where it can be relied upon to be stable. The rod has to be protected from any surrounding unstable soil. Several level stations are set into the building and their level is measured regularly with reference to the stable datum. The stations can be of the screw-in type taken away after each reading. It has been found, however, that substantial galvanized nails protruding no more than 25mm from the face of the building are not particularly vulnerable and certainly no target for vandals, and these are suitable for domestic or commercial buildings with minimal public access.

Damage is usually monitored by measuring the width of cracks (Figure G.2, *overleaf*). Usually, three measuring points are set firmly across each monitored crack, enabling its opening or closing to be measured in

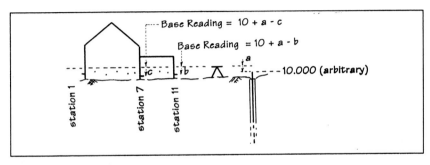

Figure G.1   Level monitoring.

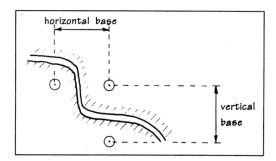

Figure G.2
Crack width monitoring.

orthogonal directions; horizontal and vertical are generally the most convenient. The measuring points may be studs or screws, and reading may be by calliper or Demec™ gauge. Alternatively, two halves of a permanent gauge may be fixed either side of each crack, enabling direct reading to be made through a vernier scale etched into the gauge.

Ideally, monitoring should include level and crack reading. The latter, especially, may on its own yield ambiguous results, because cracks can open for one reason and progress for other reasons, including thermal expansion and contraction. When level and crack width changes are studied in combination, there is seldom a doubt about cause, effect or severity.

It is also possible to monitor the lean of walls, the horizontal movement of walls, vibration, slope stability, soil moisture content – almost any feature whose potential change is of sufficient importance to justify the cost of setting up and reading frequently enough to enable a sound decision to be made on the basis of the results.

Monitoring that fails to lead to a sound decision is a waste of money, and, worse, a waste of time that might have been devoted to appropriate repairs. Successful monitoring relies on results being:

- accurate (consistent and capable of detecting changes below the threshold of damage)
- regular and frequent (so that seasonal or cyclic influences can be discerned)
- unambiguous (easily linked to the cause of movement and damage).

Clays

Overconsolidated clays are so called because after their deposition under water they were covered, and so consolidated, by a considerable thickness (overburden) of later deposits. Subsequent erosion has removed all or most of the overburden and brought the overconsolidated clay to, or near, the surface. Consolidation by overburden will have made the clay strong and stiff. In the UK, overconsolidated clays include London clay, Oxford clay, Gault clay, Kimmeridge clay and others. Virtually all overconsolidated clays are highly shrinkable and have low or very low permeability.

Boulder clay was formed during the Ice Ages by glacial action, which ground up soil and rock, producing a mixture of material that can include clay, sand, gravel, cobbles and boulders. The mixture was transported across country as the ice sheets advanced and retreated, eventually being deposited as an irregular layer, in some places thinly covering older soil or rock; in other places, filling deep valleys. Not all glaciated soil contains clay. Although extremely variable, it is usually firm (Table 3.2, *page 31*) or medium dense (Table 3.3, *page 31*). Shrinkability and permeability depend largely on clay content. Most boulder clays are less shrinkable and more permeable than most overconsolidated clays.

Further reading

The following organisations provide a wide range of updated literature which is relevant to structural repair:

Building Research Establishment (BRE)
Construction Industry Research and Information Association (CIRIA)
Health and Safety Executive (HSE)
Brick Development Association (BDA)
Arboricultural Advisory and Information Service
Society for the Protection of Ancient Buildings (SPAB)
Construction Industry Training Board (CITB)
English Heritage

In addition, certain regulations apply, or are relevant, to structural repair:

The Current Building Regulations, HMSO
National House Building Council (NHBC) Standards, issued by NHBC, Amersham
Zurich Technical Manual, issued by Zurich Municipal, Farnborough, Hampshire

BIBLIOGRAPHY

Addis, William (1990) *Structural Engineering the Nature of Theory and Design*, Chichester: Ellis Horwood Limited.
Anstey, John (1996) *Party Walls and what to do with them*, 4th edn, London: RICS Books.
Addleson, Lyall and Rice, Colin (1991) *Performance of Materials in Buildings*, Oxford: Butterworth-Heinemann Ltd.
Architectural Heritage Fund (1998) *Funds for Historic Buildings in England and Wales: A Directory of Sources*, London: The Architectural Heritage Fund.
Ashurst, John and Nicola (1988) *Practical Building Conservation Series* Volumes 1–5 Vol. 1 *Stone Masonry*; Vol. 2 *Brick, Terracotta and Earth*. Vol. 3 *Plasters, Mortars and Renders*; Vol. 4 *Metals*; Vol. 5 *Wood, Glass and Resins,* English Heritage Technical Handbook. Gower Technical Press Ltd.

Atkinson, M.F., (1993) *Structural Foundations Manual for Low-rise Buildings* London: E and FN Spon.

Barnes, G.E. (1995) *Soil Mechanics Principles and Practice*, Basingstoke: Macmillan Press Ltd.

Beckmann, Poul (1995) *Structural Aspects of Building Conservation*, Maidenhead: McGraw-Hill.

Bell, F.G. (1981) *Engineering Properties of Soils and Rocks*, London: Butterworth & Co. Ltd.

Bell, F.G. (1993) *Engineering Treatment of Soils*, London: E & FN Spon.

Biddle, Giles (1998) *Tree Root Damage to Buildings*, 2 volumes, Wantage: Willowmead Publishing Ltd.

Brereton, Christopher (1995) *Repair of Historic Buildings: Advice on Principles and Methods*, London: English Heritage.

Brohn, David (1990) *Understanding Structural Analysis*, 2nd edn, Oxford: BSP Professional Books.

Brunskill, R.W. (1990) *Brick Building in Britain*, London: Victor Gollancz Ltd.

Brunskill, R.W. (1985) *Timber Building in Britain*, London: Victor Gollancz Ltd.

Building Research Establishment (1995) *Assessment of Damage in Low-Rise Buildings with Particular Reference to Progressive Foundation Movements*. BRE Digest 251, revised 1995, BRE.

Carper, Kenneth L. (1989) *Forensic Engineering*, New York: Elsevier Science Publishing Co. Inc.

Carter, M. and Symons, M.V. (1989) *Site Investigations and Foundations Explained*, London: Pentech Press Ltd.

CIRIA Report 111 (1986) *Structural Renovation of Traditional Buildings*, London: Construction Industry Research and Information Association.

CIRIA Report 133 (1994) *A Guide to the Management of Building Refurbishment*, London: Construction Industry Research and Information Association.

CIRIA Report 166 (1997) *CDM Regulations – Work Sector Guidance for Designers*, London: Construction Industry Research and Information Association.

CIRIA Special Publication 130 (1997) *Site Safety A Handbook for Young Professionals*, 2nd edn, London: Construction Industry Research and Information Association.

Clifton-Taylor, Alec and Ireson, A.S. (1983) *English Stone Building*, London: Victor Gollancz Ltd.

Construction Industry Training Board, *Construction Site Safety – Safety Notes* (GE700), latest edn, King's Lynn: CITB.

Control Board of Finance of the Church of England (1986) *A Guide to Church Inspection and Repair*, 2nd edn, London: Church House Publishing.

Cook, Geoffrey K. and Hinks, Dr. A.J. (1992) *Appraising Building Defects Perspectives on Stability and Hygrothermal Performance*, Harlow: Longman Scientific and Technical.

Cunnington, Pamela (1984) *Care for Old Houses*, Sherborne: Alphabooks Ltd.

Cunnington, Pamela (1988) *Change of Use The Conversion of Old Buildings*, Sherborne: Alphabooks Ltd.

Cunnington, Pamela (1988) *How Old is Your House?*, Sherborne: Alphabooks.

Curtin, W.G., Shaw, G., Parkinson, G.I., Golding, J.M. (1994) *Structural Foundation Designers' Manual*, Oxford: Blackwell Science Ltd.

Davies, V.J. and Tomasin, K. (1996) *Construction Safety Handbook*, 2nd edn, London: Thomas Telford.

Department of the Environment (1996) *Design Principles of Fire Safety*, London: HMSO.

Driscoll, R. (1983) 'The Influence of Vegetation on the Swelling and Shrinkage of Clay Soils in Britain' *Géotechnique*, XXXIII No. 2, pp 93–105.

Fink, Susan (1997) *Health and Safety Law for the Construction Industry – Mason's Guide*, London: Thomas Telford.

Francis, A.J. (1989) *Introducing Structures Civil and Structural Engineering, Building and Architecture*, Chichester: Ellis Horwood Limited.

Harlow, Peter A., (Ed.) Managing Building Maintenance, (1984) Ascot: The Chartered Institute of Building.

Harris, Richard, Discovering Timber-Framed Buildings, (1978) No.242 in the Discovery Series. Aylesbury: Shire Publications Ltd.

Health and Safety Executive, *A Guide to the Health and Safety at Work etc. Act 1974 Guidance on the Act*, LI (latest edn), London: HSE Books.

Health and Safety Executive, *Occupation Exposure Limits*, EH40/current edn (regularly updated), HSE Books.

Heyman, Jacques (1995) *The Stone Skeleton*, Cambridge: Cambridge University Press.

Hinks, John and Cook, Geoff (1997) *The Technology of Building Defects*, London: E & FN Spon.

Holland, R.H., Montgomery-Smith, B.E and Moore, J.F.A. (eds) (1992) *Appraisal and Repair of Building Structures*, London: Thomas Telford Ltd.

Hollis, Malcolm (1991) *Surveying Buildings*, 3rd edn London: RICS Books.

Illston, J.M. (ed) (1994) *Construction Materials Their Nature and Behaviour*, 2nd edn, London: E & FN Spon.

Institution of Structural Engineers (1991) *Guide to Surveys and Inspections of Buildings and Similar Structures* (Nov. 1991), London: The Institution of Structural Engineers.

Institution of Structural Engineers, *Appraisal of Existing Structures*, (current edn), London: The Institution of Structural Engineers.

James, M. (ed) (1996) *Risk Management in Civil, Mechanical and Structural Engineering*, London: Thomas Telford Publishing.

Kaminetzky, Dov (1991) *Design and Construction Failures Lessons from Forensic Investigations*, USA: McGraw-Hill Inc.

Kletz, Trevor (1994) *Learning from Accidents*, 2nd edn, Oxford: Butterworth-Heinemann Ltd.

Managing Construction for Health and Safety Construction (Design and Management) Regulations 1994 Approved Code of Practice. (1995) L54, London: HMSO.

Institution of Structural Engineers (1997) *Manual for the Design of Plain Masonry in Building Structures* (July 1997), London: Institution of Structural Engineers.

Marshall, Duncan; Worthing, Derek and Heath, Roger (1998) *Understanding Housing Defects*, London: The Estates Gazette.

Maude, Thomas (1997) *Guided by a Stone-Mason The Cathedrals, Abbeys and Churches of Britain Unveiled,* London: I B Tauris Publishers.

Mays, G.C. and Smith, P.D. (eds) (1995) *Blast Effects on Buildings*, London: Thomas Telford Publications.

Melville, Ian A. and Gordon, Ian A. (1997) *The Repair and Maintenance of Houses*, 2nd edn, London: The Estates Gazette.

Melville, Ian A., Gordon, Ian A. and Boswood, Anthony (1974) *The Structural Surveys of Dwelling Houses*, 2nd edn, London: The Estates Gazette.

Mika, S.L.J. and Desch, S.C. (1988) *Structural Surveying*, 2nd edn, Basingstoke: The Macmillan Press Ltd.

Morton, J., *The Design of Laterally Loaded Walls*, (latest edn), Windsor: Brick Development Association.

Owen, Stephanie (1997) *Law for the Construction Industry*, 2nd edn, Harlow: Longman (co-published with the Chartered Institute of Building).

Ozelton, E.C., and Baird, J.A., *Timber Designers' Manual*, (latest edn), London: Crosby Lockwood Staples.

Pearson, Gordon T. (1992) *Conservation of Clay and Chalk Buildings*, Donhead St Mary: Donhead Publishing.

Petroski, Henry (1994) *Design Paradigms: Case Histories of Error and Judgement in Engineering*, Cambridge University Press.

Planning Policy Guidance Notes *Historic Buildings and Conservation Areas*, PPG 15, London: HMSO.

Planning Policy Guidance Notes, *Archaeology and Planning*, PPG16, London: HMSO.

Powell, Christopher (1984) *Discovering Cottage Architecture*, No. 275 in the Discovery Series, Aylesbury: Shire Publications Ltd.

Powell-Smith, Vincent and Billington, M.J., *The Building Regulations Explained and Illustrated*, (current edn) Oxford: Blackwell Science Ltd.

Powys, A.R. (1929 reprinted 1995) *Repair of Ancient Buildings*, SPAB.

Property Services Agency (1989) *Defects in Buildings*, London: HMSO Books.

Pyramus and Thisbe Club (1996) *The Party Wall Act Explained: A Commentary on the Party Wall etc. Act 1996*, The Pyramus and Thisbe Club.

Richardson, Barry A (1991) *Defects and Deterioration in Buildings*, London: E & FN Spon.

Richardson, Clive (June 1985) *AJ Guide to Structural Surveys*, London: The Architects' Journal.

Robertson, Bernard, Vignaux, G.A. (1995) *Interpreting Evidence Evaluating Forensic Science in the Courtroom*, Chichester: John Wiley & Sons Ltd.

Robson, Patrick (1991) *Structural Appraisal of Traditional Buildings*, Donhead St Mary: Donhead Publishing.

Seeley, Ivor H. (1987) *Building Maintenance*, 2nd edn, Basingstoke: The Macmillan Press Ltd.

Seward, Derek (1994) *Understanding Structures Analysis, Materials and Design*, Basingstoke: The Macmillan Press Ltd.

Smith, J.T. and Yates, E.M. (1987) *On the Dating of English Houses from External Evidence*, Offprint No. 59 from *Field Studies*, Field Studies Council, Williton.

Sowden, A.M. (ed) (1990) *The Maintenance of Brick and Stone Structures*, London: Chapman and Hall.

Sutherland, Stuart (1992) *Irrationality The Enemy Within*, London: Constable and Co Ltd.

Swindells, David J. and Hutchings, Malcom (1993) *A Checklist for the Structural Survey of Period Timber Framed Buildings*, London RICS Books.

Tomlinson, M. J., *Foundation Design and Construction*, (latest edn), London: Pitman.

Walker, Robert (1995) *The Cambridgeshire Guide to Historic Building Law*, Cambridge: Cambridgeshire County Council.

Waltham, A.C. (1994) *Foundations of Engineering Geology*, London: Blackie Academic and Professional.

Watt, David and Swallow, Peter (1996) *Surveying Historic Buildings*, Donhead St Mary: Donhead Publishing.

Wilson, M and Harrison, P. (1993) *Appraisal and Repair of Claddings and Fixings*, London: Thomas Telford Ltd.

Wright, Adela (1991) *Craft Techniques for Traditional Buildings*, London: B.T. Batsford Ltd.

Yeomans, David (1997) *Construction Since 1900: Materials*, London: Batsford Ltd.

York, Stephen (1996) *Practical ADR*, London: Pearson Professional Ltd.

Index

Access 229, 292
Accidental damage (buildings) 110, 246
   see also Impact .
Accidental support 19, 43, 49, 207
Adjoining owners 277
Adjudication 278
Ageing 130
Aggregates 46, 123, 127, 139, 147, 211, 216,
   219, 223
Alkali aggregate reaction 128
Allowable bearing pressure *see* Bearing
   pressure or Safe bearing capacity
Alluvium 45, 86, 93
Alteration (structural) *see* Structural
   alteration
Alternative dispute resolution (ADR)
   279–281
Alternative load path 110
Ancient monuments 54
Angle of draw 97
Anode 132
Anti-heave precautions 40–42, 92, 162,
   164–165, 169, 172, 260
Appraisal *see* Structural appraisal
Approved Code of Practice (ACOP) 58–59,
   294
Arbitration 279
Arboriculturist 256, 261
Arch 8, 42
   conversion to beam 213
   flat 206
Arching 161, 206
Archaeology 54, 250
Assessment (of material properties) 50,
   181–184, 227

Base metal 132
Bats 186
Beam
   bearing 118, 131, 182, 191, 227
   repairs 184–195
Beam action 160–161, 213
Beam and post/wall (structural form) 8
Bearing pressure (on soil) 26, 36, 41, 44, 86,
   89–90, 93, 99, 102, 175, 205, 208,
   295–296
   see also Safe bearing capacity

Bed joint profile 212
Bedding down (underpinning) 157, 272
Bedding plane 146, 172
Beetles 136–137, 182
Bending
   hogging 200–201, 214, 287
   in plane 197, 200–201, 214,
   moment 11–20, 52, 119, 123, 144, 184,
      189, 191, 231–232
   out of plane 139, 200–201, 214, 287
      see also Out of plumb
   sagging 200–201, 214, 287
Black ash (in mortar) 139
Blast 110, 227, 241, 273
Blinding 104
Bonding (masonry) 18, 143, 145, 222
Bonding timber 124, 193, 196, 282
Borascope 249
Boulder clay 30–31, 85–86, 93, 96, 299
Boulders 161, 299
Bowing 116, 143–144, 197
   see also Lean
Bracing 158–159, 187, 232, 284
Brackets 110
Brick reinforcement *see* Reinforced
   brickwork
British Standards 2, 7, 50, 184, 288
Brittle failure 13, 197
Buckling 15, 48, 111, 118, 119, 178, 227
Bulging 197, 200
Building Preservation Notice 55
Building Regulations 86, 227
Buttress 202, 206–209

Calcium aluminate 127
Calcium carbonate 146
Calcium chloride 126
Calcium silicate 127
   brickwork 70, 113–114, 116, 123, 127
Calcium sulfo–aluminate 115
Carbon dioxide 126, 146
Carbonation 106, 126
Cathode 132, 137
Cavity wall 15, 18, 115, 141, 204
CCTV surveys 242, 249
Chalk
   earth walling 147, 223

Chalk *cont.*
    soil/rock 35, 37, 83, 93, 96–97, 101, 166,
        221, 241
Change of occupancy 115, 134
Change of use 115, 246
Characteristic values (e.g. strength, applied
    load) 47, 49, 87, 183, 222, 260
Charring 121, 124, 186, 190
Chimneys 46, 109, 130, 141, 145, 203, 211
Classification (of soil) 24
Clay (soil) 21–42, 80–94, 101, 127, 157,
    257–268, 296, 299
Clay lump 147–149
Coarse soil 21–42, 80, 82–83, 94, 295
Cob 222
Cobbles 105, 299
Cohesion (of soil) 31, 36, 265–266
Collapse *see* failure (structural)
Collapse compression 83, 98, 268–269,
    295–296
    see also Inundation
Commons (bricks) 145
Compaction 34, 41, 103–104, 222, 224–226,
    295
Composite action 184–185, 187–190
Compression (force or stress) 117–118, 139,
    178, 184, 189
Compulsory purchase 55
Concealed defects 71, 284
    see also Hidden details
Condensation
    interstitial 128, 134–135, 139, 143, 191
    surface 130, 134, 139, 182
Connections 5, 50, 132, 137, 154, 184, 187,
    189, 192, 227–235, 247, 252, 255
    see also Fixings and Joints
Conservation 2, 53–55, 124, 155, 181, 186,
    202, 211, 217, 239, 247, 250
Conservation areas 54, 276
Consolidation (masonry core deterioration)
    144, 217, 272
Consolidation (of soil) 25–26, 34, 80–82, 84,
    94, 98–99, 104, 208, 267, 295–296
Consolidation (masonry repair) 203
Construction (Design and Management)
    Regulations (CDM) 58–61, 243, 276,
    278, 281
Contamination 35, 127, 167, 176, 283
    see also Pollution
Continuous strip foundations 41
Continuous strip underpinning *see*
    Underpinning: continuous strip
Contract (for construction work) 155, 159,
    253, 273
Contract law 275–276, 278
Contraction (of materials) 112–118,
    120–121, 154

Conversion (HAC) 125–126
Copings 130, 134, 143
Corbels 110
Core (masonry) *see* Rubble core (masonry)
Corrosion 131–132, 137, 139, 140–141,
    179, 187, 204
Cost 53, 55, 59, 68, 159–160, 179, 203, 219,
    240–243, 245, 250, 252–255, 258–260,
    266, 270, 279, 282, 284, 292
Creep
    brickwork 118, 138, 217
    downhill movement of soil 101
    roof spread 191, 206
    settlement 80–82, 98, 171
    timber 19, 118, 119, 189, 191
Critical tensile strength (in masonry) 15, 117
Cross walls 15, 148, 198, 207
Cryptoflorescence 146
Crystallization 146

Damage
    BRE categories 197–199
    cosmetic 25, 89, 155, 198–199, 272
    discovery 257
    incipient 282
    onset 256, 263
    progressive 117, 154, 270, 282
        see also Movement: progressive
    serviceability 198–199
    stability 198–199
    threshold 85, 268, 272, 298
Damage indicators *see* Indicators
Damp 106, 128, 130, 134–135, 141, 148,
    179, 182, 191, 221, 250, 252, 272, 283,
    287
    see also Condensation, Rain penetration or
        Rising damp
Dead load *see* Load:self weight
Deathwatch beetle 182, 187
Debris (after fire) 120–122, 247
Deflection *see* Distortion
Deformation *see* Distortion
Defrassing 183
Delamination 145–146, 201, 214, 217, 249
Demolition 121, 127, 245, 269
Dendrochronology 286
Dereliction 246
Desiccation 86, 90
Destructive testing or investigation 183, 202,
    248, 284
Deterioration 95, 102, 107, 129–149,
    153–154, 202–206, 210–212, 221–222,
    225, 251–252
Diagnosis 2, 51, 65–66, 70, 102–103, 198,
    202, 256, 270–271, 282
Differential movement 33, 93, 171, 255, 266,
    297

Differential pressure (soil) 89–90
Differential settlement 33–34, 41, 102, 171, 208, 255, 297
Differential subsidence 297
Distortion 9, 13, 18–19, 46, 51–52, 90, 99, 108–109, 118, 121, 139, 145, 154, 172, 187, 198, 200, 208, 210, 243, 250, 266, 283–284, 287–288
    see also Strain
Distortion survey 200, 233, 248–249, 256–257, 262–263, 270–271
Distribution curve 47–51, 87
Disturbing forces 38–39
Double shear 232
Dowels (in joints) 107, 162, 193, 196
Downdrag 42
Drainage materials 97, 171, 242, 268
Drawdown (by pumping) 94
Dry rot 124, 130, 133–138, 186–187
Ductile failure 13, 197
Durability (of repair) 147, 180, 210, 212, 217, 223, 229
Dust explosion 241

Earth walling 124, 147–149, 221–226, 235
Efflorescence 146, 219
Electrolytic action 131, 137, 187
Electrolytic series 132
Emergency repairs 55
    see also Urgent rescue
End grain 138
Endoscope 249
English Heritage 55, 276
Epoxy mortar 219
Epoxy resin 190–196
Equilibrium
    soil moisture 26–31
    structural 9–20, 99, 108–111, 121, 153, 189, 200–201, 223, 287
Erosion (of masonry) 139, 146
Erosion (of soil) 82–83, 97, 200, 241, 267–268, 296
Escape of water 39, 82, 99, 241
Ettringite 127
Evaporation (of rain) 146, 210
Evaporation (from soil) 24, 85, 105
Event tree 292
Expanded metal reinforcement 219–220
Expansion (of materials) 112–118, 120–121, 154, 174, 210
Explosion *see* Blast

Facade retention 246, 254–255
Facings (bricks) 145
Failure (structural)
    brittle 13, 197
    ductile 13, 197

    gradual 197, 287
    incipient 100, 247
    sudden 118, 197–198, 287
Falsework 247
    see also Temporary support
Fasteners *see* Connections
Fatal accidents 56, 59–60
Feasibility 246, 252–253
Ferrous oxide 131
Fill *see* Made ground
Fillers 176, 190, 203, 216
Fine soil 21– 42, 80, 295
Fire 120–124, 181, 185, 189, 195, 209–210, 246–247, 252, 273, 292
Fixings 50, 192, 208, 228–235
    see also Connections and Joints
Flint 221–222
Flitch plates 186–195
Flood 273
Floor
    ground bearing 103–107, 174–179
    in situ concrete 103–107, 174–179
    precast concrete 106, 179
    suspended 106, 174–175, 234
    timber 107, 178
Forces
    bending *see* Bending
    direct (tension or compression) 9–20, 46, 52, 123, 231
    secondary 49, 227, 233
    shear *see* Shear force
Foundation types 40–42
Frame
    medieval 6, 12, 51, 71, 108, 114, 180, 191, 195, 242, 282
    modern 5, 8, 191, 195, 232, 286
Friction (in soil) 31, 36, 83, 295
Frost 130, 141–143, 146, 148

Gabion wall 270
Gang nailed gussets 187
Geological fault 35, 97
Geotechnical engineer 39–41, 101, 167, 265
Girder 8, 187
Glaciation 35, 101, 299
Gradual failure *see* failure:gradual
Granite 35, 37, 44, 113
Granular soil *see* Coarse soil
Gravel 21–42, 81, 94, 105, 299
Gravel rejects 105
Grit blasting 147
Ground anchor 39
Ground beam 41, 163, 167
Ground penetrating radar 249
Ground slab *see* Floor
Ground water 35, 83, 93, 96–97, 101, 105, 127, 141, 166, 176, 292

Ground water lowering 93–95, 296
Ground water rise 82–83, 93, 97, 268, 292
Grout injection 174–179, 190, 213–217, 223
Grout pressure 216–217
Gusset 187

Hanger 14, 52, 232
Hardcore 103–105, 174–178
Hardwood 113, 180
Hazards (to buildings) 7, 33, 41, 75–79, 98, 157, 252, 265
Hazards (to humans) 35, 58, 61–64, 147, 157, 162, 241–244, 253, 292
Health and Safety at Work etc. Act 57, 59, 276
Health and Safety Commission (HSC) 58
Health and Safety Executive (HSE) 58, 292, 294
Heartwood 138
Heave
    clay 25–31, 42, 83–93, 155, 161, 201, 263–268, 270, 272
    frost 32, 93
    lateral 42, 91–92, 266
    shale (in hardcore) 105, 174
Hidden defects see Hidden details or Concealed defects
Hidden details 71, 229, 254, 282
    see also Concealed defects
High alumina cement (HAC) 125–126
Historic buildings 53–55, 83, 93, 139, 186, 218, 247, 250–251, 262, 276
Hogging see Bending:hogging
Hygroscopic salts 130, 203

Impact 71, 110, 148, 247, 273
    see also Accidental damage
Incipient damage see Damage:incipient
Incipient failure see Failure:incipient
Indicators (of cause of damage) 66–70, 92, 103, 200–202, 256–257, 263, 271
Industrial floors 107, 179
Infestation see Beetles
Infill panels 51, 108, 184, 282
Injection see Grout injection
Insects see Beetles
Inspection of work 60, 65, 254, 276
Insurance claims 2, 256–273, 278–279
Integrity testing (CFA piles) 166
Intervention 2, 53, 245, 248, 255
Inundation 82–84, 99, 268, 273, 295
    see also Collapse compression
Investigation 2, 35, 70, 81, 99, 107, 120, 159, 204, 246–253, 269, 271, 273, 276, 283, 288
Iron pyrites 128

Jacking 171–173, 187, 189, 202, 208–209, 272
Joints (timber) 114, 187–196
    see also Connections and Fixings
Joist trimmers 110
Joists 15, 19, 108–109, 185, 192, 232, 283
Junctions (of walls) 17, 118, 121, 198, 250

King post 52

Landslide/landslip 36–42, 100, 265, 269–270
Larvae 137
Lateral heave see Heave: lateral
Lateral movement see Movement: lateral
Leakage see Escape of water
Lean (of walls) 46, 100, 118, 119, 172, 197–202, 207
    see also Bowing
Level survey see Distortion survey
Life expectancy (materials) 129, 204
Like for like 54, 187–189, 211, 213
Likelihood ratio 67, 69
Lime (in earth wall repair) 223
Lime (in mortar)
    see Mortar containing lime
Lime wash 218
Limestone 37, 96, 113, 123, 146–147, 210
Lintels 149, 190, 206, 213
Liquid limit 24, 27, 29
Listed Building Consent 53, 55
Litigation 273, 274–276, 279
Load
    characteristic see Characteristic values
    eccentric 144, 197, 202, 208, 228
    horizontal 5, 7, 11, 17, 232
    imposed 11, 19, 47, 50, 125, 179, 180, 184, 189, 206
    lateral 208, 228, 233
    paths 8, 17, 46, 109–110, 181–182, 213, 228, 232
    self-weight 5, 11, 110, 148, 184, 189, 206, 208, 228
    testing 51, 68, 106, 166, 183, 250
    thermal 121
    uniformly distributed 11
    wind 5, 46–47 see also Wind suction
Local planning authority 54–55, 93, 218, 247, 250, 253–254, 271, 276
Longhorn beetle 182
Loss adjuster 273

Machinery vibration see Vibration: machinery
Made ground 34, 36, 82–84, 95, 98–99, 161, 241, 267–268, 295–296
Magnesian limestone 147

Magnesium sulfate 147
Map cracking 128
Margin of safety 12, 38–39, 99, 138–139,
    142, 145, 157, 175, 202, 222, 265, 272
Masonry bee 130
Mean strength 47–48
    see also Characteristic values
Medieval timber frame see Frame: medieval
Metal detector 249
Middle third 16
Mini-piles see Piles: mini
Mining
    deep 42, 97
    shafts 97
    shallow 97
Mining waste 83
Mitigation (of structural damage) 90,
    154–155, 161, 214, 219, 260, 268–269
Model (of structure) 6, 10, 19, 43–52, 144,
    183, 250, 252
Modulus of elasticity 13
Moisture content see Water content
Moisture movement (in clay) 24
    see also Permeability: clay
Moment see bending
Monitoring 68, 81, 89, 93, 98, 118, 124,
    133, 155, 168, 181, 191, 246, 251–252,
    262, 267–273, 297–298
Mortar containing cement 114–115, 127,
    138, 141, 145, 211
Mortar containing lime 123, 141, 210–211
Mortar (general) 46, 113, 123, 139, 141,
    145, 233
Mortar (hard) 142, 212
Moulds (for shaping timber repairs) 190,
    193, 196
Moulds (for shaping masonry repairs) 203
Movement
    cyclic 18, 112, 117, 119, 153–154, 298
    differential see Differential movement or
        Differential settlement
    free 113, 121, 255
    gross 209
    lateral 99, 118, 255
    of support 182
    out of plane 139, 198, 200, 214, 232
    progressive 80, 85–89, 119, 139, 153–155,
        206, 258
        see also Damage: progressive
    repetitive 118, 144
    seasonal 84–86, 90, 154, 170, 258, 298
    thermal 112–113, 117, 119, 121, 154,
        201, 210
Movement joints 114, 117, 154
Mudstone 37
Mundic 128

Nail sickness 137
Natural frequency 95
Negative pore pressure 84
    see also Suction (in clay)
Negligence 274, 278
NHBC guidelines (regarding trees) 86–87,
    260, 262
Noble metal 132
Non destructive testing or investigation 202,
    248–250
Notching (timber) 109

Obsolescence 1, 246
Ombudsman (insurance) 273, 278–279
Optimum density 104
Options for repair 1, 115, 153–156, 161,
    186–187, 193, 203, 241, 252, 258, 260,
    267, 270, 272
Ordinary Portland Cement (OPC) 126, 223
Organic soil 33–34, 104
Out of plumb 197, 201, 223
    see also Bending:out of plane
Overburden 38, 93, 100, 299
Overconsolidated clay 31, 85–86, 93, 127,
    299
Overload 111, 118, 120, 182, 187, 205, 227,
    243, 247
Oxidation (of metal) 131
Oxidation (of peat) 34

Pad 16, 41, 163, 205
Parapets 134, 141, 203, 211
Partial safety factor 48
Particle size 22, 24, 32, 176, 223
Party Wall etc. Act 277
Passive pressure 91
Peat 33–34, 45, 93, 95
Penetrating damp see Rain penetration
Permeability
    mortar 142, 210
    render 142, 219
    soil 24, 26, 32, 39, 80, 84–86, 92–95, 258,
        263, 299
Persistent moisture deficit 84
Piezometer 93
Piles 161, 164–166, 272
    augered 166
    CFA 166–167
    driven 166–167
    mini 167, 174, 177–178
    sheet 39, 270
    sleeved 165
Pisé 222
Planning consent 53
Planning supervisor 58
Plaster 147, 282

Plastic limit 24
Plasticity 24–33, 223, 262
Plasticity index 24
Pointing 130, 142
    see also repointing
Pollution 35, 127, 146, 283
    see also Contamination
Pore water pressure 26–31, 39, 84–85, 167, 295
Pre-loading 172
Prestressing 161, 208
Preventive action (maintenance or repair) 2, 99, 133, 154, 239–244, 261
Professional indemnity 278, 281
Progressive movement see Movement: progressive
Propping 121, 159, 167, 181, 187, 189–190, 195, 247
Pruning 261–263
Pulse radar 249
PFA (as filler) 176
Pumping 32, 93–95
Purlin 19, 119, 183

Radius of influence (pumping) 93–95
Radon 35
Rain penetration 128, 130, 134, 142–144, 182, 191–192, 212, 218, 221, 242, 250, 283
    see also Damp
Rat runs 225
Ratchet effect 117
Records 160, 181, 192, 239–240, 247–248, 251, 286
Recovery (of soil after subsidence) 27, 84–90, 161, 258–264, 272
Redeposited soil see Made ground
Redistribution (of stress, load or bearing pressure) 16, 44–46, 86, 99, 118, 138, 168, 183
Refurbishment 2, 7, 51–53, 245–249
Reinforced brickwork 155, 192, 196, 204, 206–207, 213–214, 218, 220, 260
Reinforced concrete 41, 107, 123–127, 155, 160–171, 191, 205, 207
Reinforced plaster 155, 192, 218–220, 260
Reinforced render 155, 218, 260
Relative humidity 131
Relocation 173
Removal of cause 156, 256–271
Render 116, 147–148, 209, 217–219, 222–223, 282
Render (hard) 142, 146, 149, 222
Render (reinforced) see Reinforced render
Repairs Notice 55
Repair only option 154, 258, 260, 266–269

Repointing 203, 210–213
    see also Pointing
Resin 186, 193, 195, 205, 214–215, 220
Resistivity 249
Restoring forces 39, 101
Restraint 17, 109, 112, 116–117, 121, 140–141, 145, 154, 157, 178, 185, 200, 207–208, 222–223, 229–235, 255
Retaining wall 99, 141
Reversible repairs 186, 194
Rising damp 106, 148, 179
    see also Damp
Risk assessment 61, 87, 101, 111, 184, 240–241, 261–262, 272, 292
Risk management 61, 241, 261–262, 280–281
Robustness 17, 51, 110, 118, 154, 157, 227–229, 232, 282
Roof spread 119, 191, 200–201, 206
Roof sway 201
Root barriers 161, 261, 265
Root growth 25, 29, 86–87, 91, 101, 105, 260–263, 271
Root pruning 261
Rot 110, 120, 124, 133–138, 143, 178, 186, 191–192, 229, 252, 273, 287–288
    see also Dry rot or Wet rot
Rubble core (masonry) 18, 143–144, 214–217
Rust 106, 126–127, 131, 141, 204

Safe bearing capacity (of soil) 36
    see also Bearing pressure
Safety factor 48
Safety margin see Margin of safety
Sagging see Bending:sagging
Sand 21–42, 83, 94–95, 223, 241, 292, 299
Sand–lime brickwork see Calcium silicate brickwork
Sandstone 37, 113, 123, 146–147, 210, 221
Sapwood 138
Saturation (of clay) 27–29
Schmidt hammer 177
Screed 103, 126, 174, 176–179
Seasonal movement see Movement: seasonal
Security 247, 252
    see also Vandals
Sensitivity 96, 149, 262, 292
Serpula lacrymans see Dry rot
Serviceability see Damage
Settlement 26, 34, 80–82, 102, 103–105, 107, 109, 153, 173, 174, 176, 197, 209, 296
    see also Differential settlement
Shaft friction (piles) 272
Shale 37, 105

Sharp sand 211, 223
Sheet piles 39, 270
Shear force 11–12, 197–198, 231, 287–288
Shoring 121, 158–159, 181, 190, 247, 292
Shrinkability (of soil) 24, 33, 93, 299
Shrinkage
    clay 25–31, 84, 91, 262, 296
    drying 114–115, 120, 124, 188, 211, 219,
        225, 273
    earth wall 223
    limit (of clay) 27, 29
Sidesway 118, 119
Silica 128, 147
Silicosis 147
Silt 21–42, 83, 93, 95, 166, 241, 292
Single incident (causing damage) 153
Single pile support 164
Site rules 253, 276
Six metre rule 277
Slab see Floor
Slate 37
Slenderness 15, 18, 109, 139, 140–141, 145,
        148, 167, 178, 180, 185, 202, 207, 222,
        227
Slenderness ratio 15–16, 140
Slope
    destabilizing 38–39, 100, 265–266
    instability 36–41, 100, 266
    stabilizing 38–39, 101
Snap headers 144
So far as is reasonably practicable 58
Sodium chloride 126
Soft rock 221
Softwood 113, 180
Soil layers 32, 94
Soil mixtures 22, 32, 94
Soil/structure interaction 44–46, 89–90
Sole plate 41, 138, 188, 242
Solution (of rock) 96
Spalling 109, 120–124, 143, 209
Splits (in timber) 114, 191, 195
Spot listing 55
Stabilizer 223–224
Staged tree removal 263–264
Steel
    austenitic 131
    stainless 131, 191, 203, 229
Stiffness (of building and materials) 12,
        45–46, 51, 89, 109, 141, 180–183, 189,
        229
Stiffness (of soil) 24, 89, 102
Stitching 192, 195, 213–217
Strain 12, 46, 97, 189, 197
    see also Distortion
Straps 90, 155, 159–161, 184, 191, 213–214,
        231, 235, 269

Straw (in earth walling) 147, 223
Strength (of building and materials) 12– 20,
        45–46, 50, 90, 121–124, 139, 143,
        148–149, 180–183, 191, 202, 210, 217,
        219, 222–223, 229, 251
Strength (of soil) 24, 31–41, 154
Stress 12–20, 46, 117, 121, 123, 138, 145,
        180, 233
Strip footing 41
Structural alteration 44, 108, 138, 145,
        148–149, 182, 206, 243, 282, 284, 296
Structural appraisal 15, 44, 51, 139, 181,
        187, 246, 282
Structural engineer 43, 46–47, 95, 107
Structural form 8
Strut 14, 52, 183, 284
Stud and mud 148
Sub-base 103, 105, 107
Subgrade 103–107, 179
Subsidence 25, 34, 82–89, 93–98, 197, 201,
        241, 256–263, 267–272, 296
Suction (in clay) 24–31, 84–89, 296
Suction testing (in clay) 30
Sudden failure see Failure: sudden
Sulfate attack 105, 115, 127, 174, 203, 219
Sulfur dioxide 137, 146
Surface depression (after subsidence) 96–98
Surface depression (in earth walling)
        224–225
Surveyor 135, 284
Suspended ground floor see Floor: suspended
Swallow holes 96, 241
Swelling (of clay) 25, 90–91, 161

Tail (of distribution curve) 48, 87, 260
Task difficulties 186, 204, 229, 235, 254
Temperature
    fire 121–122, 190
    normal variation 18, 101, 117, 133–135,
        141, 233
Temporary accommodation 253
Temporary support 109, 167, 181–182, 187,
        225, 243, 247, 252, 254
    see also Falsework
Tension (force or stress) 117–118, 139, 144,
        179, 184
Thaumasite 127
Thermal movement see Movement: thermal
    or Temperature: normal variation
Thermal shock 120
Thermography 225, 249
Thixotropy 190, 215
Three metre rule 277
Tie 160–161, 191–192, 195, 233, 284
Tilting see Lean (of walls)
Timber frame see Frame

Timber treatment 135, 137, 181, 183, 186–187
Timing of repairs 272
Tort 274
Transition (in underpinning) 170–171
Transpiration 24, 85, 105
Tree Preservation Order (TPO) 271
Tree removal *see* Vegetation control
Tree root action *see* Root growth or Vegetation control
Tree root identification 262, 271
Trench fill 41
Trial holes 217, 250, 270–271, 278
Tricalcium aluminate 115
Truss 8, 11, 52, 183, 187
Tunnelling 98

Uncertainty 1, 19, 43–52, 87, 155, 260, 262, 272, 280
Underpinning 81, 161–171, 202, 259, 263, 265, 267–269, 271
    continuous strip 161–162, 170, 292
    pad and beam 161, 163, 170
    partial 168, 170–171
    pile and beam 161, 164, 170
    piled raft 165
Unforeseen ground conditions 159–160, 280
Urgent rescue 54, 139, 246–247, 259, 267
    see also Emergency repairs

Vandals 247, 252
    see also Security
Variations (contract) 159–160, 254, 282
Vegetation control 28–29, 39, 87, 171, 257–266, 271
Vent (in stonework) 146
Ventilation 128, 130, 178, 192, 252
Venting (explosive forces) 110, 241
Venting (fire) 120
Vermin 130, 148, 221
Verticality measurements *see* Distortion survey
Vibration
    airborne 95–96
    groundborne 32, 95–96
    machinery 95–96
    traffic 149
Viscosity (repair materials) 190, 215–216
Volume change (of clay) 25–32, 34, 83, 93, 295

Wall cavity 139–141, 229
Wall ties 139–141, 204–205
Water bearing seams (in soil) 166
Water content (of soil) 23–39, 92, 105, 266
Water demand 84, 86, 261–262

Water shrinkage factor 30
Water table 84, 93–95
Wear and tear 130, 179, 221
Weatherproofing 247, 252
Wet rot 124, 133–138, 186
Wind suction 201

Young's modulus 13